I0014611

Stylianos Papanastasiou

Investigating TCP Performance in Mobile Ad Hoc Networks

Stylianos Papanastasiou

Investigating TCP Performance in Mobile Ad Hoc Networks

VDM Verlag Dr. Müller

Imprint

Bibliographic information by the German National Library: The German National Library lists this publication at the German National Bibliography; detailed bibliographic information is available on the Internet at http://dnb.d-nb.de.

Any brand names and product names mentioned in this book are subject to trademark, brand or patent protection and are trademarks or registered trademarks of their respective holders. The use of brand names, product names, common names, trade names, product descriptions etc. even without a particular marking in this works is in no way to be construed to mean that such names may be regarded as unrestricted in respect of trademark and brand protection legislation and could thus be used by anyone.

Cover image: www.purestockx.com

Publisher:
VDM Verlag Dr. Müller Aktiengesellschaft & Co. KG, Dudweiler Landstr. 125 a, 66123 Saarbrücken, Germany,
Phone +49 681 9100-698, Fax +49 681 9100-988,
Email: info@vdm-verlag.de

Copyright © 2008 VDM Verlag Dr. Müller Aktiengesellschaft & Co. KG and licensors
All rights reserved. Saarbrücken 2008

Produced in USA and UK by:
Lightning Source Inc., La Vergne, Tennessee, USA
Lightning Source UK Ltd., Milton Keynes, UK
BookSurge LLC, 5341 Dorchester Road, Suite 16, North Charleston, SC 29418, USA

ISBN: 978-3-639-03035-8

Acknowledgements

I would like to thank my supervisors Dr. M. Ould-Khaoua and Dr. L. M. Mackenzie for their support and guidance through the course of this research. Although strict at times, their criticism and acute observations have been pivotal in improving this work and achieving meaningful deliverables.

I must further extend my gratitude to my colleagues and staff in the Department of Computing Science at the University of Glasgow for their help and advice. In particular, I would especially like to thank the people I have collaborated and co-authored research papers with. In no particular order these are Muneer M. Bani Yassein, Vassilis Charissis and Shaliza Wahab.

Also, eternal gratitude is owed to my family who have been supportive in everything I have ever done. In particular, I would like to thank my father, Vasilios for his never-ending patience and understanding. Also, I am highly and forever indebted to my mother, Euterpi, for her unlimited moral support and for having much faith in me. Finally, I owe gratitude to my brother Tasos for his acute commentary and unsolicited feedback on any non-research related activity of mine. Without my family's continuous encouragement and tolerance this work would not have been finished.

Contents

List of Tables

List of Figures

ix

Acronyms used in this dissertation

ACK   ACKnowledgement

ADV   Adaptive Distance Vector (routing)

AODV   *Ad hoc* On-Demand Distance Vector (routing)

BER   Bit Error Rate

BSD   Berkeley Software Distribution

CTS   Clear-to-Send

CWL   Congestion Window Limit

cwnd   congestion window

DAA   Dynamic Adaptive Acknowledgements

DSDV   Dynamic Destination-Sequenced Distance-Vector Routing

DSR   Dynamic Source Routing

dupACK   duplicate ACKnowledgement

FTP   File Transfer Protocol

IETF   Internet Engineering Task Force

LL   Link Layer

MAC   Media Access Control

MANET Mobile *Ad hoc* NETwork

MPR   MultiPoint Relay

OLSR   Optimised Link State Routing

PDA   Personal Digital Assistant

RFC   Request For Comments

RERR   Route ERRor

RREP   Route REPly

RREQ   Route REQuest

RTO   Retransmission TimeOut

RTS   Request-to-Send

RTT   Round Trip Time

RTTVAR Round Trip Time VARiance

SACK   Selective ACKnowledgements

SCA   Slow Congestion Avoidance

SNR   Signal-to-Noise Ratio

SRTT   Smoothed Round Trip Time

SS   Slow Slow start

SSA   Signal Stability based Adaptive (routing)

SSCA   Slow Slow start and Congestion Avoidance

TC   Topology Control

TCP   Transmission Control Protocol

UDP   User Datagram Protocol

Chapter 1

Introduction

Wireless communications have experienced explosive growth in recent years due to the wide availability and rapid deployment of wireless transceivers in a variety of computing devices such as PDAs, laptop and desktop computers. The *de facto* adoption of the popular IEEE 802.11 [52] standard has further fuelled these developments by ensuring interoperability among vendors thereby aiding the technology's market penetration. Initially, the deployment of these wireless technological advances came in the form of an extension to the fixed LAN infrastructure model as detailed in the 802.11 standard. Therein a wireless client is associated with an access point which acts as a router and arbiter between the mobile client and the rest of the network, which may include several other mobile agents, forming a Basic Service Set [52]. In contrast to wired LANs, the mobile client is not physically constrained by cables and there are even provisions for a seamless hand-off process for clients roaming in areas covered by cooperating access points, thereby ensuring extended wireless coverage. The latter configuration is referred to as the Extended Service Set in IEEE 802.11 nomenclature [52].

As the processing power and transceiver capabilities of mobile clients increased, it became feasible to use the clients themselves as forwarding agents. In particular, instead of using fixed infrastructure in the form of access points, the mobile nodes may cooperate in a peer-to-peer fashion to forward each other's messages. By acting as routers, willing hosts may form the backbone of a spontaneous network which facilitates connectivity and services for interested parties. The term Mobile *Ad hoc* NETwork (or MANET for short) has been coined [91] to describe such a network and

1

the concept has proven significant enough for the IETF to form a working group on the subject [80]. In fact, the IEEE 802.11 standard itself makes provisions for a rudimentary *ad hoc* mode of operation between stations when an access point is not present, in the form of the Independent Basic Service Set [52]. This defines the presence of a communications link between two parties without the need of an access point to coordinate and forward transmissions. However, such a configuration is only applicable to stations within mutual communications range and requires the cooperation of higher level protocols for the formation of multihop paths.

The potential significance of MANETs lies in the promise of ubiquitous connectivity provided that mobile hosts can communicate effectively, given the special constraints of the hosts themselves as well as the unique dynamic topological characteristics of the formed network. Particular applications of MANETs include scenarios where infrastructure is expensive to set up and difficult or even impossible to deploy, such as battlefield or disaster relief operations [58]. Other uses include plugging "holes" in the coverage of wireless infrastructure [23] or even integration with cellular 3G+ networks [9] to achieve wider connectivity.

1.1 MANET characteristics

Mobile *ad hoc* networks share many of the properties of wired-infrastructure LANs but also possesses certain unique features which derive from the nature of the wireless medium and the distributed function of the medium access mechanism. These constraints may be described in turn as considerations stemming from the *wireless channel*, the *mobile node* and the *routing protocol* used to establish and maintain communication paths. These characteristics affect the functionality of mechanisms throughout the communication protocol stack and are considered now in turn.

1.1.1 Channel characteristics

Signal attenuation: As the transmitted signal spreads out from the aerial in all directions it attenuates as distance increases. As such, the intensity of the electromagnetic energy at the receiver decreases with distance from the transmission; beyond a certain distance, the signal-to-noise ratio (SNR) becomes so low that the receiver is not to able

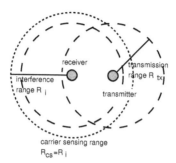

Figure 1.1. The transmission, carrier sensing and interference ranges in a communicating pair

to decode the transmission successfully.

Taking the above into account and for an omni-directional transceiver, three ranges may be identified [103] as shown in Figure 1.1. These are, from the sender's perspective:

Transmission Range (R_{tx}) The range within which a transmitted frame can be successfully received by the intended receiver. Within this range the SNR is high enough for a frame to be decoded by the receiver.

Carrier Sensing Range (R_{cs}) The range within which the transmitter triggers carrier sense detection. When this happens, the medium is considered busy and the sensing node defers transmission.

Interference Range (R_i) The range within which an intended receiver may be subject to interference from an unrelated transmission, thereby suffering a loss. This range largely depends on the distance between the sender and the interfering node.

Those ranges are related to one another, with $R_{tx} \leq R_i \leq R_{cs}$, as the energy required for a signal to be decoded is greater than what is needed to cause interference [63]. The interference and transmission ranges depend on the signal propagation model and the sensitivity of the receiver, assuming that power constraints apply and all transmitters transmit at the maximum allowed power level [103].

Multipath fading: Multipath fading occurs because of different versions of the same signal arriving at different times at the receiver. These versions effectively follow different paths from the transmitter, with different propagation delays, due to multiple reflections off intervening obstacles. The superposition of these randomly phased components can make the multipath phenomenon a real problem especially if there are many reflective surfaces in the environment and the receiver is situated in a fringe area of reception [64].

Transmission errors: Due to the volatile nature of wireless signal propagation, the wireless medium potentially exhibits errors. The frame format in IEEE 802.11 networks is similar to the 802.3 (Ethernet) format [98] and uses the same 48-bit MAC address fields. The specification also includes the IEEE 802.3 32-bit CRC polynomial-based error detection mechanism. However, the protection offered by this scheme only extends to data actually travelling on point-to-point links (as opposed to end-to-end). It is still possible, though somewhat unlikely, for corrupt data to be accepted by the receiver, but this is offset by the adoption of error discovery, implemented at higher layers (such as the checksum fields in IP, TCP and UDP segments).

Hidden and exposed terminals: Consider the scenario illustrated in Figure 1.2(a). Node A is transmitting to node B. Node C cannot receive the transmission and since its carrier sense function detects an idle medium, it will not defer transmission to D and a collision will be produced at node B. In this case, node A is hidden with respect to node C (and vice versa). This problem is offset in 802.11 by using a short packet exchange of Request-to-Send (RTS), Clear-to-Send (CTS) frames. This is a two-way handshake where the source terminal transmits the RTS to the destination which then replies with a CTS frame. If there is no reply, then transmission is deferred as presumably the medium at the area around the destination is busy. If a CTS reply is received then DATA transmission follows. Since the duration of the transmission is included in the RTS/CTS exchange, neighbouring nodes defer their transmissions for the time the medium is occupied. Point-to-point transmission is reliable since the DATA frame is followed by an ACK transmission from the destination if the frame is successfully received.

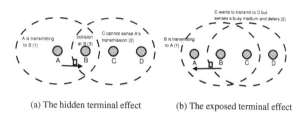

(a) The hidden terminal effect   (b) The exposed terminal effect

Figure 1.2. Illustration of the hidden and exposed terminal effects

The exposed terminal effect occurs when a station that needs to transmit a message senses a busy medium and defers transmission even though it would not have interfered with the other sender's transmission. An instance of the exposed terminal effect is demonstrated in Figure 1.2(b). Here, node B is transmitting to node A. Node C senses node B's signal and defers transmission. However, it need not have done so as C's transmission does not reach node A and would not have interfered with B's transmission at the location of the intended destination (node A). Node C is the exposed terminal in this case. Note that both the hidden and exposed terminal effects are related to the transmission range. As the transmission range increases, the hidden terminal effect becomes less prominent because the sensing range increases. Nonetheless, the exposed terminal effect then becomes more prominent as a greater area is "reserved" for each transmission.

In the above examples, the transmission (R_{tx}), interference (R_i) and carrier sense (R_{cs}) ranges are all assumed to be equal. However, several research efforts have concentrated on the effects of interference on the hidden and exposed terminal effects [25, 26, 104, 106, 107] when $R_i > R_{tx}$. In particular, Xu et al. [103] have shown that when the distance d between the source and destination nodes is $0.56 * R_{tx} \leq d \leq R_{tx}$, where R_{tx} is the transmission range of the sender, the effectiveness of the RTS/CTS exchange declines rapidly.

Spatial contention and reuse: Network links among hosts are commonly fixed in wired networks and do not interact with each other as there is typically little interference between physical cables. In contrast, wireless network links operate differently.

Assuming omnidirectional antennas, when a node transmits, it "reserves" the area around it for the transmission's duration; i.e. no other transmission is to take place during that time interval as it will result in a collision and waste of bandwidth. Spatial reuse refers to the number of concurrent transmissions that may occur in a network without interfering with each other. It is the responsibility of the MAC protocol to ensure that transmissions are coordinated in such a way so as to maximise the property of spatial reuse.

For illustration purposes consider that in Figure 1.3 communication between nodes $0 \rightarrow 1$ and $4 \rightarrow 5$ may happen simultaneously. Then, communication among other node pairs could happen concurrently in turn, as long as each pair is 4 hops apart from the other. Since at most two pairs can transmit at the same time without affecting each other, the spatial reuse of this string topology is 2. It should be noted that the spatial reuse in a particular scenario represents an optimal level of concurrency; it is not always achievable and it may be the case that with enough nodes transmitting simultaneously packets will be lost due to interference. Such a situation is referred to as *spatial contention* and it can become the main cause of packet drops when a path is long enough as noted in [43]. This is in contrast with wired networks where packet drops are mainly caused by buffer overflows at the routers.

Capture effect (interplay of TCP with 802.11 MAC):   In wired networks, TCP has a well documented bias against long (as in hop length) flows [39]. In 802.11 multihop networks, the bias is much stronger and is manifested in the form of the *channel capture effect*. Essentially, if two TCP connections are located in near vicinity of each other and, thus, interfere with one another, this effect favours the session that originated earlier or the one that flows over fewer hops. The favoured session often starves the other almost completely with data transport not being accomplished for the mistreated session until the other one has completed all of its data transmission.

The bias is rooted in the exponential backoff of the Distributed Coordination Function of the 802.11 MAC mechanism, which is inherently unfair and is further augmented by TCP's own exponential backoff mechanism. The string topology scenario as shown in Figure 1.3 aptly illustrates the point. Even though the two TCP flows do not share the same path, flow 2 may starve because of flow 1; unfairness may also be present when the flows share some or much of the same path [69]. The phenomenon

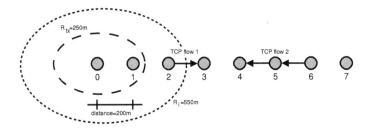

Figure 1.3. String topology setup where the capture effect may occur

has become a focal point of research interest and has been extensively explored in the literature [69, 104, 107].

1.1.2   Characteristics of MANET nodes

Mobile nodes that participate in a MANET operate under limitations which have to be considered in networking protocol design if new proposals are to be efficiently operable in such environments. As participating nodes may be heterogeneous in nature (laptops, PDAs or even desktop systems), to ensure mobility and some degree of autonomous operation, devices often have limited power reserves and possibly limited processing capabilities [61]. These restrictions are typically discussed in the literature [29] in the context of proposals for new routing algorithms or service-providing mechanisms over MANETs and are briefly outlined below.

In general, energy is a scarce and valuable commodity for MANET nodes and its consumption can therefore be just as important a measure as throughput, latency and other traditional performance metrics when evaluating MANET protocols at any layer. Several methods have been proposed to conserve energy at various levels, including the operating system and applications. An overview of approaches to power conservation through energy-aware mechanisms is included in [61]. Specifically, from a TCP perspective, power savings are best achieved by minimising redundant retransmissions whenever possible [73]. The savings in this case are twofold; the source conserves power by transmitting fewer packets but also every forwarding node in the path benefits since fewer unnecessary retransmissions occur.

Although high speed wireless communications are possible [53], it is assumed that MANETs are primarily characterised by relatively bandwidth-constrained wireless links [29] compared to their hardwired counterparts and, furthermore, the capacity of such links is variable. In particular, after taking into consideration the effects of interference, multipath fading and so on, as presented in Section 1.1, the transmission rate of a mobile node may be severely affected. The design of any level in the protocol stack should take account of this constraint by minimising overhead where possible and any proposed mechanism should be usable across possibly asymmetric links of different capacities.

Another important node characteristic is the potentially restricted CPU capacity at each node. Routing algorithms in particular are designed to be simple so as to operate with little processing and storage requirements [92]. It follows that any adjustments proposed to TCP, or indeed any other networking protocol, should minimise complexity, so that CPU time costs do not outweigh gains in other metrics (e.g throughput or latency). In addition, heavy CPU usage requires more power which makes processor-intensive modifications even more costly. Surprisingly, although power considerations are a key focus of routing and clustering techniques in MANETs, proposed modifications are not always examined with respect to their overall power and CPU demands [61].

Finally, security is a salient concern in MANETs as in most forms of wireless communications. As messages are exchanged through a common transmission medium, it becomes difficult to prevent snooping on network traffic. Consequently, security provisions notwithstanding, the network is vulnerable against replay attacks, eavesdropping and message redirection, even more so than in the case of a wired infrastructure. Security measures normally applicable to wired LANs are also largely applicable to MANETs, although special precautions are necessary in the case of routing as balancing the security overhead and limited available bandwidth is still a matter of ongoing research [88].

1.1.3 Routing in MANETs

MANETs are potentially characterised by significant node mobility which induces highly dynamic topologies and may even result in partitioned networks. In this section, network partitioning and the effects of routing failures are discussed in addition to current general approaches to routing protocol design in such multihop environments.

Network partitioning: Network partitioning occurs when, due to mobility, nodes which were able to communicate directly or through the cooperation of other nodes at some time, T_1, are unable to do so at a later time, T_2, because there is no longer a usable path between them. It is further possible that at a still later time, T_3, the nodes have placed themselves in such a position that the network is again connected and every node can reach every other one, either directly or indirectly. The scenario is illustrated in Figure 1.4.

TCP is not engineered to deal with network partitioning as it is not normally a frequent occurrence in wired networks. In effect, the exponential backoff of TCP's retransmission timeout (RTO) mechanism[1] facilitates the exponentially delayed probing of a valid path. This may be illustrated in the topology depicted in Figure 1.4(a) where the TCP source (node 1) is communicating with the destination (node 6) at time T_1 through nodes [2-5]. Then, in Figure 1.4(b) at time T_2, node 4 has travelled outwith the range of node 3 and their mutual link has become invalid; there is a network partitioning with two separate and isolated network partitions, namely A and B. Packets in this time frame do not get forwarded to the destination and ACKs do not reach the source as there is no usable path available. The TCP agent then enters an RTO phase (due to packet loss) as the underlying routing protocol attempts to discover an alternate route. A new TCP segment is sent every time the RTO timer expires. If this segment reaches the destination, TCP continues with normal transmission and the route is utilised. However, since the RTO is doubled after every timeout, those "probing" packet transmissions take longer each time. Hence if the disconnection persists for consecutive RTOs, there might be long periods of inactivity during which the network may be connected again, but TCP is still in the backoff state [34]. As a result, throughput suffers.

Routing failures: In MANETs, unlike wired networks, route failures occur frequently due to the mobile nature of the participating nodes. Route failures may also

---

[1] A detailed outline of TCP operations is included in Chapter 2

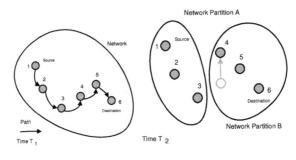

(a) An *ad hoc* network   (b) Network partitioning

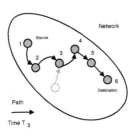

(c) Network is connected again

Figure 1.4. Illustration of network partitioning

occur when repeated point-to-point transmissions fail, for instance because of the effects of spatial contention [106]. When a link failure occurs, the routing protocol attempts to discover an alternate path, but the duration of the restoration period largely depends on the mobility of the nodes, the mechanism of the routing protocol itself and the network traffic characteristics.

The effects of route failures on TCP operation resemble those of network partitioning as discussed previously. If the route takes some time to restore, TCP enters its backoff state and sends "probes" for a restored route at increasingly longer time intervals. Hence, the route might be restored for quite some time but TCP remains idle until it launches the retransmitted packet after RTO expiration and receives a reply from the destination. Further complications arise from TCP's round trip time (RTT) calculation. After the route has been re-established the RTT measurements required for the RTO calculation should reflect the characteristics of the new route. However, since the RTO is calculated using a weighted average of new and old RTT measurements, for some time after the route restoration, the RTO value contains a mixed estimate of the old and new route characteristics. The above concern applies in wired networks as well but may be more severe in MANETs, where route reconstructions are expected to be a frequent occurrence [16].

Routing approaches in MANETs: Due to the dynamic and unpredictable topologies of MANETs, proposed routing algorithms have to operate by introducing little overhead, be light-weight enough to meet the mobile node constraints and still be able to forward packets efficiently to their designated destinations. Further, measures ensuring that packets are free from looping behaviour should be present as route loops can occur frequently in such a continuously changing topological landscape. Two main approaches to routing in MANETs currently exist and are now outlined in turn.

Proactive Routing This approach involves the proactive discovery of paths to potential destinations by preemptively building a view of the network. As soon as a node is introduced in the network it attempts to discover as much as it can about the topology around it, i.e. discover ways to route packets to different destinations, in anticipation of future communication attempts. Note that as routes to all possible destinations are maintained at all times there is no initial delay when setting up a connection.

The traditional Bellman-Ford [12] algorithm, as used in wired networks [47] has been shown not to be efficient in a MANET environment [93]. Special considerations are needed to ensure loop freedom such as including a monotonically increasing sequence number in route updates, a technique adopted by several contemporary approaches [60, 92]. However, the overheads of table exchanges in proactive routing become increasingly a point of concern as the network topology expands and/or grows more dynamic.

The Optimised Link State Routing (OLSR) mechanism [28] is a popular proactive routing protocol that has advanced through the procedures of the IETF to become a proposed RFC standard. This introduces several optimisations to minimise the excessive overheads associated with proactive routing whilst implementing a table driven route exchange mechanism. The node still maintains tables to multiple routes but special nodes called multi point relays (MPRs) are chosen in the network and become the focal points of organisation so as to minimise broadcast and route discovery/maintenance overheads.

Reactive routing In traditional hardwired networks routing failures are infrequent and hence the cost of route discovery is negligible. The communication links are not expected to change radically through time and so a fixed initial cost in terms of throughput and delay as routes are being discovered is presumed negligible. In multihop wireless networks with highly dynamic topologies (such as MANETs), however, this cost can increase substantially if route breakages are frequent enough, as the route discovery cost is incurred multiple times.

To mitigate this overhead, it may be preferable for the routing agent to maintain and discover routes only on an as-needed basis. As such, alterations to link status that are of no interest to a node's current communications operations are not taken into account, leading to significant overhead savings in the case of frequent route changes. The downside of such an approach is the initial delay when discovering a route for a new connection as well as the relatively high overhead of route discovery which requires a network-wide broadcast in an effort to contact the desired destination. There are two notable reactive routing protocols progressing through the IETF; *Ad hoc* On Demand Distance Vector (AODV) [92] and Dynamic Source Routing (DSR) [60]. Several research efforts have been reported in the literature in an attempt to keep the cost of route discovery at a reasonable level [21, 28].

There are also mixed/hybrid approaches which combine proactive route discovery and maintenance for nodes within a given hop radius distance with a reactive approach [82] for the rest of the network. The proactive/reactive dichotomy is a fundamental routing design decision; the IETF has acknowledged both approaches as having merit and is maintaining proposed experimental RFCs for both [80].

1.2   Related work

The vast majority of TCP evaluation studies over MANETs have been carried out with simulations [2, 6, 24, 25, 31, 34, 42, 43, 49, 65, 70] and a few with limited experimental testbeds [7]. The complex interplay of TCP with the routing protocol and the wireless access mechanisms makes the development of analytic models of TCP behaviour extremely difficult and, it would be fair to say, an aspiration as yet unmet.

Ahuja et al. [2] have conducted the first evaluation of TCP performance under different routing algorithms over MANETs, namely the *Ad hoc* On-Demand Distance Vector (AODV) [92], the Dynamic Source Routing (DSR) [60], the Dynamic Destination-Sequenced Distance-Vector Routing (DSDV) [93] and the Signal Stability based Adaptive (SSA) [33] routing protocols. This work has shown the detrimental effect of mobility on TCP Tahoe as well as the relative merits of SSA routing which incorporates signal strength as a measure of path optimality instead of hop distance.

An investigation of TCP Reno over AODV, DSR and the Adaptive Distance Vector (ADV) [68] routing protocols has been performed by Dyer et al. [34]. The ADV routing protocol has been shown to maximise TCP throughput, with AODV being the second best performer under various mobility conditions. Further, the throughput penalty of utilising stale cache entries under moderate mobility has been noted in the case of DSR. The authors have also proposed the use of a heuristic named the *fixed-RTO*, which greatly improves TCP throughput in AODV and DSR, as it aids TCP in utilising restored routes quickly without resorting to feedback from the routing protocol. In the same work, however, the authors stress that this solution is MANET-oriented and not intended for use when there exists a gateway to other networks (such as the Internet).

The effects of interference on TCP, as noted in Section 1.1, have also been widely studied in the literature [6, 25, 31, 43, 65, 70]. Xu et al. [106] have examined the throughput of TCP Tahoe, Reno, NewReno, SACK and Vegas over multihop string topologies in an open space environment under the 802.11 protocol. The authors have demonstrated that TCP Tahoe and its variants (Reno, NewReno and SACK) exhibit throughput instability in such topologies, as interference causes packet drops which are interpreted as congestion losses. TCP Vegas does not suffer from this problem due to its conservative mechanism for increasing the sending rate. Further investigation has revealed that other TCP variants may regain throughput stability by limiting the maximum congestion window (i.e. the maximum number of packets in flight) to four segments. The viability of using the delayed acknowledgement option has been discussed in the same work and shown to improve throughput by 15-32% in the same static topologies.

The work in [106] has been complemented through further study by Fu et al. [43]. Here, the authors have noted that for string, cross and mesh topologies there is an optimal congestion window (*cwnd*) size which maximises throughput by improving spatial reuse, i.e. which facilitates the maximum possible non-conflicting simultaneous transmissions. It has been further noted that since TCP continuously increases its *cwnd* size until packet loss is detected, it typically grows and operates at an average window size that is larger than optimal, thereby causing spatial contention. The authors have also obtained the optimal window size for each of the above mentioned topologies. Finally, they have proposed two link layer modifications named "link RED" and "adaptive pacing" to aid optimal spatial reuse and improve TCP performance.

The issue of TCP throughput instability and effective spatial reuse has moreover been addressed in [25, 26]. It has been shown that by making general MAC layer assumptions, i.e. not 802.11 specific, the bandwidth-delay product in multihop MANET paths cannot exceed the product of the Round-Trip Hop Count (RTHC) of the path and the packet size. In the case of the 802.11 protocol this bound is shown to reach no more than a fifth of the RTHC. According to this adapted definition of the bandwidth delay product, the authors have then proposed an adaptive mechanism which sets the maximum cwnd according to the route hop count, noting an 8-16% throughput improvement. In [26] the performance merit of TCP-pacing, which evenly spaces a window's worth of packets over the current estimated round-trip time, has also been evaluated but no worthwhile performance improvement was found.

A number modifications to the TCP receiver for optimal spatial reuse have been proposed in the literature [6, 31] and are largely complementary to sender-side modifications. The proposed alterations invariably include some type of management of acknowledgement (ACK) transmissions. Such a line of enquiry is similar to work undertaken in the past for wired networks [3] but whose premises and conclusions are not directly applicable to MANETs. In particular, work by Altman et al. [6] has revealed that the rate of ACK production in TCP can affect TCP throughput across long enough paths. It has been demonstrated therein that by making aggressive use of delayed cumulative ACKs, the number of ACK responses produced, which vie for transmission time with TCP data, may be reduced. This facilitates more efficient spatial reuse as fewer packets compete along the path for point-to-point transmissions. The authors have also proposed a new delayed ACK scheme which stretches the TCP ACK-clocking property (one ACK response per at most two segment receptions) and results in throughput improvements of 9-22% in string topologies when small data transfers occur. Further, Oliveira et al. [31] have suggested an alternative end-to-end technique that combines constraining the sender's congestion window and implementing a dynamic ACK delay window at the receiver. In this approach, the receiver delays an ACK response for 2-4 segments received but also maintains a dynamic timer which depends on segment inter-arrival time and which triggers an immediate ACK response upon expiration. Subsequent evaluation in static topologies has revealed reduced retransmissions and, in the case of multiple flows, a compelling throughput improvement of up to 50% over regular TCP. A different approach to the subject which forgoes the end-to-end modification constraint has been undertaken in work by Yuki et al. [112], where DATA and ACK segments are combined in intermediate nodes' transmissions so as not to occupy two separate transmission slots; it is then up to their respective destinations to identify the portion of the combined packet addressed to them.

Anastasi et al. [7] have conducted measurements on an actual testbed and have verified simulation results by showing that interference becomes a serious problem in *ad hoc* networks when TCP traffic is considered. Moreover, the authors have noted that during their experiments there was sufficiently high variability in channel conditions at different times to make comparison of results difficult. They have also observed that certain aspects of real wireless transmissions are not effectively captured in simulation. These include the different sending rates of the preamble, the RTS/CTS and the DATA frames in 802.11 networks, as well as the variability of the transmission and physical sensing ranges (even within the same session). Plesse et al. [94] have conducted real life experiments with military scenarios in urban areas using OLSR and have confirmed TCP throughput problems as the hop count of the path increases, as well as the transient quality of the signal during trial runs. In the same work, it is noted that the RTS/CTS mechanism is not a prerequisite to achieving good spatial reuse in their 2-hop path experiments, an observation in concert with earlier simulation results presented by Xu et al. [103].

1.3 Motivation

A number of previous research studies have dealt with the behaviour of TCP in MANETs [2,35,74] and highlighted relevant problems. Some of these enquiries have focused on the differences between routing protocols and have made use of a single type of TCP agent to measure performance discrepancies amongst those protocols [2,34,49]. These types of studies typically vary the topology conditions in terms of route breakage frequency and network partitioning time to simulate the degree of mobility present. Such attempts are of particular significance as it is not enough to utilise UDP (congestion unaware) traffic to fully evaluate the effectiveness of routing protocols in MANETs [108]. Further, since most TCP variants share similar congestion avoidance mechanisms, an observation particular to a specific variant is likely to be applicable to others although potential applicability is usually not explicitly investigated.

Other enquiries [10, 24, 46, 66, 77, 104] have focused on the particular interaction of TCP agents with a selected MANET routing protocol taking additional account of some general MANET trait (e.g. packet loss due to errors, or mobility). The potential shortcoming of such approaches is that as routing protocols may vary significantly (for instance consider on-demand vs proactive routing), resulting observations may not be applicable to different routing agents. Further, since the potential TCP shortcoming is dealt with in isolation [26], it is not immediately apparent whether an improvement may be seen in a more general setting. In a number of studies [26,95], such optimisations are applied to static TCP scenarios which may offer significant improvements in special cases but whose performance merits may be reduced in a more typical MANET scenario.

A study among TCP variants and, even more specifically, among reactive and proactive TCP agents has been lacking in the literature. Further, the absence of selective acknowledgement (SACK) enabled TCP agents in the different performance evaluation efforts is of particular note as these are widely deployed and implemented in modern operating systems [4]. As a result it has long been an open research question whether the type of TCP agents proposed in the literature are competent and efficient performers in multihop wireless environments.

A crucial observation on the fitness of the 802.11 standard as a basis for MANET infrastructure has been made by Xu and Saadawi [107]. The authors have discovered fundamental throughput related inefficiencies in the widely adopted IEEE 802.11 MAC mechanism when used for multi-hop communications. The authors have further noted the severity of these performance reducing effects particularly when TCP is in use for end-to-end communications and have proposed a solution by modifying a simple TCP parameter (the maximum *cwnd*). Several researchers have expanded on the work in [107] by either applying changes in the MAC layer, and so potentially breaking 802.11 compatibility [43], or by keeping the MAC layer intact but using intra- and inter-layer communication to smooth out the MAC shortcomings [104]. Fu et al. [104] have shown that across long paths, packet drops due to MAC issues are overwhelmingly more numerous than buffer overflow induced losses, which are the main cause of packet loss in wired networks. The original end-to-end modification suggested by Xu et al. [106] and subsequent follow up work [43], have been initially evaluated in static topologies so as to study the effect in isolation, precluding the effects of mobility. As such the interplay of proposed solutions with other factors involved in projected mobile MANET scenarios, namely the effects of mobility and mobility-induced packet loss, have not been taken into consideration.

Finally, most work on end-to-end approaches to reduce the number of packets from individual connections on the same path has focused on sender-side restrictions [25]. These have the advantage of effectively reducing the problem because the source is the main contributor to spatial contention, as TCP DATA segments are typically much larger than ACKs and require substantially greater transmission times. Nonetheless, recent efforts have concentrated on reducing the number of ACKs produced by the

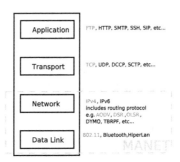

Figure 1.5. A protocol stack diagram highlighting the protocols considered in this dissertation

receiver, which are vying for transmission time and competing with DATA packets [6, 31].

Evaluations of such techniques have been inadequate in certain aspects. Firstly, not all options at the MAC layer have been exploited even though the 802.11 standard has defined multiple modes of operation; in particular disabling the RTS/CTS mechanism during short frame transfers, has not been considered. Further, since substantial progress has been made on the simulation tools and MANET systems have been progressively better understood, conclusions reached in previous, early research may have actually been reflections of peculiarities of the simulation tools used, or side-effects of misplaced simulation assumptions and may not necessarily be representative of future MANET systems [75, 110].

1.4 Thesis Statement

The goals set for this dissertation derive from the motivations as listed in the previous section and may be summarised in the following thesis statement:

The aim of this work is to propose new methods to ease the negative effects of TCP's misinterpretation of the causes of packet loss in MANETs. The changes introduced should be as simple as possible with respect to implementation complexity and should not break the end-to-end TCP paradigm.

The scope of this work may be better understood by considering Figure 1.5; the diagram included therein depicts the layers in a typical MANET protocol stack and highlights in red the particular protocols this dissertation deals with.

1.5 Contributions

To address the above research concerns this dissertation undertakes an extensive performance comparison between TCP variants and presents two new mechanisms to improve spatial reuse and maximise TCP throughput in MANETs.

The first part of this dissertation contains a thorough and detailed analysis of simulation traces of the basic TCP Reno agent in a controlled route-breakage scenario under three popular routing protocols, namely AODV, DSR and OLSR. This demonstrates the complex interaction of the TCP agent with the routing protocol under the conditions of route breakage in MANETs. An extensive overview of the artifacts introduced by this interaction is presented and subsequently used to highlight the problems of traditional TCP agents in MANETs with regards to non-congestion related losses. Further, extensive simulation results of the performance of three reactive (Reno, NewReno and SACK) and one proactive (Vegas) TCP variants are viewed in light of different routing protocols. These results reveal the difference in performance between the variants across routing mechanisms and are accompanied by a detailed account tracing the causes of this discrepancy. Notably, the merits of each variant are explained and evaluated in the context of dynamic topologies. Further, TCP Vegas is shown to be decisively competitive, throughput-wise, with regard to the modern reactive TCP agents, NewReno and SACK in MANETs.

The second part of the dissertation introduces a novel sender-side technique, inspired by the performance merits of the TCP Vegas congestion avoidance mechanism, which results in better spatial reuse over standard 802.11 transceivers. As such the method mitigates the effects of spatial contention caused by MAC-layer mis-coordination. The new technique is derived by considering modifications to TCP's congestion avoidance without compromising its fundamental principle of additive increase/multiplicative decrease. Subsequent performance analysis of the enhanced agent is conducted using both static and dynamic topologies and the new proposal is compared and contrasted with an existing solution from the literature. The resulting discussion places both solutions in context of static and dynamic MANET scenarios discussing their relative merits in each situation.

Finally, this dissertation also deals with receiver side modifications intended to achieve better spatial reuse for the TCP agent. Although there has been some work in the literature on the subject, careful examination of simulation traces, as well as experience drawn from past MANET simulation research, reveals several shortcomings and omissions in both the solutions themselves and their evaluation. Drawing on those lessons of the past, and through extensive simulation, a more thorough evaluation of past approaches is conducted and, in addition, a new technique is presented which improves performance through careful setting of MAC parameters, without compromising the assumption of an 802.11 MAC mechanism. The dissertation emphasises issues with existing evaluation techniques, widely used in literature, and aims to place future performance evaluation of TCP in MANETs in the context of more varied scenarios, offering suggestions for future research work.

1.6 Outline of the dissertation

The rest of the dissertation is organised as follows. Chapter 2 introduces in detail the reactive and proactive TCP variants under examination, namely TCP Reno, NewReno, SACK and Vegas. Further, a description of the routing protocols used in the simulations and subsequently relevant to the follow-up discussion are presented. Then, a list of common assumptions concerning the subsequent simulation analysis is also included. Finally, justification is offered on the method of study used in this dissertation and a rationale is given on the choice of simulation tools employed.

Chapter 3 contains a simulation comparison of the aforementioned TCP variants in a variety of mobility scenarios and under different routing protocols. Moreover, detailed simulation traces representative of the throughput behaviour of each transport protocol are included per routing protocol and their differences are highlighted.

Chapter 4 discusses TCP's role in spatial reuse in MANETs. In particular, a new approach which leads to goodput improvement through a sending rate reduction is discussed and analysed through simulation. The new technique is then evaluated against an existing end-to-end modification and its relative merits are discussed and highlighted.

Chapter 5 presents an overview of techniques that reduce the rate of ACK responses in MANETs to improve spatial reuse and outlines a new approach which uses widely implemented features of the wireless transceiver to improve performance. A discussion on the limitations of existing evaluation techniques is also included and the new results are presented in a well-specified problem domain and context.

Finally, Chapter 6 summarises the results presented in this study and offers suggestions for future research work.

Chapter 2

Preliminaries

The main objectives of this chapter are to provide some background on the characteristics of TCP agents, offer an overview of MANET routing protocols and present the mobility model and common simulation assumptions used in this dissertation. As such, the chapter is organised as follows. Section 2.1 describes the TCP agents and congestion avoidance mechanisms this dissertation deals with. Section 2.2 contains a succinct description of the main operations of the MANET routing protocols used in subsequent chapters. Section 2.3 includes a description of the random waypoint model, which is used in this work to simulate topological changes. Section 2.4 lists the common simulation assumptions which apply throughout this dissertation. Finally, Section 2.5 provides justification on the method of study and techniques used in this dissertation.

2.1 Fundamental TCP principles

The Transmission Control Protocol (TCP) [15] is a widely used transport protocol in wired and wireless communications, layered on top of IP networks to provide reliable end-to-end congestion control. Apart from establishing, maintaining and dissolving connections between communicating pairs, a TCP agent is responsible for behaving fairly towards other network flows including other TCP agents whilst not exceeding network capacity. The way this fairness and sensible resource usage is achieved though is not explicitly specified; as such there are different TCP variants, each of which nevertheless obeys basic behavioural rules [97].

TCP sends data in segments which do not exceed a *maximum segment size* as negotiated via a three-way handshake between the communicating agents during an initial connection establishment phase. Each byte (octet) of data has a sequence number assigned to it. When the receiver receives a segment, it notes the bytes of data (or sequence number range) of the segment and responds by sending back a cumulative acknowledgement (ACK) which confirms that all bytes up to the given sequence number have successfully arrived. The TCP sender also maintains a retransmission timeout (RTO) timer, which on expiration indicates that a segment has been lost and is to be retransmitted. The functionality offered by cumulative ACKs, the RTO timer as well as a checksum on the segment header and data ensures reliability on top of IP.

Another important functionality of TCP is flow and congestion control through the use of a "sliding window" [98], measured in bytes. The sending rate is throttled by the congestion window maintained at the sender and the receiving window advertised by the receiver. The minimum of the two defines the maximum amount of outstanding (unacknowledged) data that the TCP agent may maintain at any one time in the network and along the communications path for a particular connection. The adjustment of the receiving window allows the receiver to set the rate of incoming segments so that it is not overwhelmed by the load. On the other hand, tweaking the congestion window is a means for the sender to adjust to varying network conditions and avoid causing congestion in the network.

Recent traffic monitoring over the Internet [4] has confirmed the popularity of the Reno [56] and NewReno [38] TCP variants as well as the increasing adoption of the TCP selective acknowledgements (SACK) modification [14]. A promising reactive solution to the problem of congestion control has further been presented in [16] with the introduction of TCP Vegas which has received much attention in the literature [1, 45,72,109]. These TCP variants are viewed as likely candidates for adoption as reliable transport protocols for use over MANETs as they are readily implementable and safe to use over small or large scale networks. The basic principles of the aforementioned protocols, which form the focal point of this dissertation follow.

2.1.1 TCP Reno

TCP Reno refers to the implementation of the TCP protocol in the 4.3 Berkeley Software Distribution (BSD) which includes the additive increase, multiplicative decrease congestion control algorithm proposed in [56]. Congestion control is accomplished using four distinct mechanisms, namely *slow start, congestion avoidance, fast retransmit* and *fast recovery.*

The slow start phase is activated immediately after the initial handshake that establishes the connection or following the expiration of the retransmission timer. Every time an ACK is received the congestion window (*cwnd*) increases by one segment size and so effectively per round trip time (RTT), *cwnd* is doubled (i.e. increases exponentially). Initially, the slow start mechanism increases the *cwnd* until a a congestion indication event is triggered or the maximum sending rate is reached. A congestion indication event could either be three duplicate ACKs (dupACKs) or a retransmission timeout (RTO).

The slow start threshold (*ssthresh*) state variable stores the value of half the sending rate (*cwnd* size) at which the last congestion indication event occurred. The congestion avoidance phase is triggered when the *cwnd* reaches the *ssthresh* value during slow start or after the fast retransmit/fast recovery phase. During the congestion avoidance phase, *cwnd* increases linearly and up to one full sized segment per RTT. This phase attempts to gently feed segments into the network after reaching half the rate when the previous segment delivery failure occurred.

Finally, the fast retransmit/fast recovery phase occurs when the sender receives three dupACKs which indicate that a TCP segment has been lost in flight. A dupACK is sent by the receiver whenever it cannot acknowledge an arriving segment because it has not received all the segments sent prior to that one. The fast retransmit algorithm requires an immediate retransmission of the missing segment without waiting for the RTO timer to expire. Fast recovery sets $ssthresh \leftarrow \frac{1}{2} * cwnd$ and sets $cwnd \leftarrow ssthresh + (3 * \text{max.segment size})$. Then, the *cwnd* 'inflates' for each additional dupACK received so that it is possible to continue sending segments in an attempt to keep the network 'pipe' utilised while waiting for an ACK to acknowledge new data. When such an ACK arrives *cwnd* 'deflates' to *ssthresh* and TCP enters the

congestion avoidance phase. The linear increase of the sending rate (during the congestion avoidance phase) as well as its radical decrease (after an RTO or three dupACKs)
form the additive increase/multiplicative decrease property of TCP which maintains
fairness between connections sharing the link, ensures fast convergence to a fair share
state when other flows need to utilise the available bandwidth and guards against the
possibility of congestion collapse [56].

The RTO timer in TCP Reno is computed by measuring the RTTs of transmit-
ted segments. In particular the RTO value is set to $RTO \leftarrow SRTT + max(G, 4 * RTTVAR)$ every time an RTT sample is collected, where SRTT is a smoothed average of the RTT samples and RTTVAR denotes the RTT variance. TCP features an
RTO "heartbeat" counter which checks for RTO expiration at time intervals of given
length G; the interval length defines the RTO timer's granularity and is set by default
in several implementations to 100, 200 or 500ms [11, 17, 76]. The collection of RTT
samples and the reset of the RTO timer are activated on a per-window basis.

2.1.2 TCP NewReno

The NewReno TCP variant [38] improves upon the congestion recovery mechanism of
Reno without requiring changes to TCP receivers or the TCP segment format. More
specifically, the existence of a selective ACK receiver is not assumed (as opposed to
TCP SACK described in Section 2.1.3). The NewReno algorithm is functionally very
similar to TCP Reno. The difference between the two variants can be distilled to the
treatment of a loss event during the congestion avoidance phase. When the TCP agent
enters the fast recovery phase, provisions are made so that the sender actually responds
to ACKs that do "cover" new data but not all the outstanding data in the pipe at the
time the loss was detected. These are labelled partial ACKs, and in the NewReno
paradigm they prompt the retransmission of the first unacknowledged segment by the
partial ACK and the reset of the retransmission timer. This enables the TCP sender to
recover from multiple packet losses in a single window of data without resorting to the
coarse grained RTO, which is often detrimental to throughput.

Two TCP NewReno variants are specified in the relevant RFC [38], namely the
"Careful" and the "Less Careful" agents. The difference between the two is that due
to the absence of exact information on the receiver's buffer the "Careful" variant times

out if a packet preceding three other packets in flight in a window of data is lost, whilst the "Less Careful" one does not. Essentially, there is a chance that the "Less Careful" variant will fast retransmit unnecessarily on occasion but can also fast retransmit as desired in cases where the "Careful" variant has to rely on RTO expiration. The NewReno RFC [38] suggests the "Careful" variant be implemented, as it might be more conservative and at times sub-optimal but does not overburden the network with spurious traffic in any case, unlike the "Less Careful" agent. The "Careful" variant is evaluated in the simulation experiments included in this dissertation.

It is notable that the NewReno modification to TCP was widely deployed in most modern operating systems [38] long before it was ratified as an IETF standard. Its popularity may be attributed to its effectiveness in avoiding extensive RTO periods by intelligently filling gaps in the receiving buffer caused by dropped or reordered packets.

2.1.3 TCP SACK

TCP Sack [14] is a Reno-based TCP variant which makes use of the facilities provided by the Selective Acknowledgements (SACK) option of TCP [81]. In this fashion, SACK-enabled segments provide the TCP sender with some indication of the status of the destination's receiving buffer. To achieve this, the data receiver generates SACK information for every ACK response it produces that does not cover the highest sequence number in the data receiver's queue.

Hence, when reception of a non-contiguous segment occurs, instead of returning a dupACK, the receiver produces a reply which contains further information in the header of the segment in the form of an option. The information embedded on the SACK response contains a list (in the form of block pairs) of some of the isolated data blocks in the receiver's buffer, which have not been passed on to the application layer, as additional data segments are required to plug-in the gaps in the receiving sequence within the receiver's window. Hence, in the event of packet loss the sender can re-send only the exact packets that have been lost in transit and avoid producing spurious retransmissions.

The TCP SACK Option is widely supported on the Internet; even though not all agents make use of the SACK information, many produce SACK-enabled responses

[14]. There is also an optional TCP D-SACK [41] mechanism (where D stands for Duplicate) that makes use of Selective ACKs to help the sender infer the order of packets received at the destination and thus realise when it unnecessarily retransmits a packet after a duplicate ACK has been produced. This allows the sender to act more intelligently in the case of persistent reordering, packet replication or early RTOs.

2.1.4   TCP Vegas

TCP Vegas [16] introduced several new mechanisms to TCP including a proactive congestion avoidance technique which does not violate the congestion avoidance paradigm of TCP. Instead of increasing the sending rate until a segment loss occurs, TCP Vegas tries to prevent such losses by decreasing the sending rate when it senses incipient congestion even if there is no indication of segment loss. As such TCP Vegas can be classified as a "proactive" variant as opposed to "reactive" Reno-based agents which respond to segment losses after they have occurred. An overview of the main mechanisms of Vegas follows.

First, there is a different retransmission mechanism compared to TCP Reno. TCP Vegas features two timeout values. The first is the regular coarse-grained RTO value similar to the one in TCP Reno, which is limited in accuracy by the granularity of the "heartbeat" counter. The other is a fine-grained RTO value based on a more accurate RTT estimate. Both the fine and coarse-grained counters are calculated as in TCP Reno. The more accurate RTO estimate is possible because TCP Vegas measures the RTT for every segment transmitted within the sending window by reading the system clock at the segment's departure and then once more at arrival of the corresponding ACK. Consequently, a more fine-grained RTO value is calculated per RTT, which is only triggered, however, by the arrival of corresponding ACKs. Whenever a dupACK is received TCP Vegas checks whether the difference between the current time and the timestamp recorded for the relevant segment is greater than the fine-grained RTO. If this is the case, that segment is retransmitted immediately without waiting for further duplicate ACKs. Further note that for the first or second segment (depending on how many segments are in transit) after the fast retransmission there is a fine-grained RTO expiration check even on non-dupACKs.

Conversely, TCP Reno waits for 3 dupACKs before retransmission and so if enough dupACKs are lost on the way, and this is likely in case of congestion, Reno will fall back on its coarse-grained timeout mechanism. Although the TCP Vegas retransmission mechanism can be activated with a single dupACK, it is not necessarily more aggressive than Reno's as any retransmission is in accordance with the TCP specification; the RTO timer for the retransmitted segment has expired and so a retransmission is standards compliant. Since the fine-grained RTO counter is only examined when a dupACK is received, TCP Vegas may have to fall back on the coarse-grained RTO timer, like Reno, if dupACKs do not trigger the fine-grained timer. Should multiple congestion indication events occur, TCP Vegas reduces the congestion window only for the first fast retransmission as it tries is to avoid decreasing the sending rate for congestion that was observed before the last window decrease. As such, Vegas does not penalise the connection by further reducing the window's new size for effects that may be attributed to the window's previous size. The trade-off in the case of this added fine-grained timer is additional computational and storage demands on the TCP agent but these do not appear to be significant [17].

The proactive congestion control behaviour of Vegas is based on RTT measurements. Once per RTT, Vegas computes the current (actual) measured throughput and compares it with what it considers to be the expected throughput. The expected throughput is computed as $expected = \frac{windowsize}{baseRTT}$, where the $baseRTT$ is the smallest observed RTT measurement for the connection, and $windowsize$ is the number of bytes currently in flight. The actual throughput is computed as $actual = \frac{rttLen}{RTT}$, where RTT is the average RTT of the segments acknowledged during the last RTT, whilst $rttLen$ is the number of bytes transmitted during the last RTT. The difference $(diff)$ between the two measurements is calculated in $baseRTT$ segments as follows:

$$diff = \left(\frac{windowsize}{baseRTT} - \frac{rttLen}{RTT} \right) * baseRTT \qquad (2.1)$$

If the difference is under a certain threshold, α, then the congestion window increases by a full segment size since there is evidence that the expected throughput is achievable and so the sending rate should increase. If the difference is above a (possibly different) threshold, β, then this is taken as a sign of incipient congestion and the congestion window decreases by a full segment size. Otherwise, the congestion window $(cwnd)$ remains unchanged. The decision process used to adjust the sending rate per RTT is summarised below:

$$cwnd = \begin{cases} cwnd + 1 & \text{if } diff < \alpha \\ cwnd & \text{if } \alpha \leq diff \leq \beta \\ cwnd - 1 & \text{if } diff > \beta \end{cases} \qquad (2.2)$$

In the original Vegas papers [16, 17], the α and β thresholds were set to 1 and 3 respectively.

Finally, the slow start mechanism of Vegas uses a variation of its congestion avoidance mechanism to decide when to switch to the congestion avoidance phase. Vegas monitors the expected and actual rate per RTT and increases the congestion window only every other RTT to make the comparisons valid. As soon as a queue buildup is detected (i.e. $diff > 1$), Vegas moves on to the congestion avoidance phase. Vegas is much more conservative during the slow start phase compared to Reno-based variants and such behaviour is indicative of the "proactive" philosophy of Vegas. Whilst TCP Reno initially tries to fill in the pipe with enough segments to cause segment loss and thus probe the capacity of the bottleneck link, TCP Vegas chooses to measure the pipe's reaction to added segments in order to realise the available capacity without inducing segment loss.

2.2 Routing principles in MANETs

There are two main routing approaches in MANETs as expressed in IETF recommendations through the RFC process, namely the *reactive* and *proactive* routing paradigms. The former category includes the AODV [92] and DSR [60] protocols, whilst the latter is represented by OLSR [28]. The main functionality of each of these is now presented in turn.

2.2.1 *Ad hoc* On-Demand Distance Vector (AODV) routing

The AODV routing algorithm is a popular reactive routing algorithm which has been ratified by the IETF in an experimental RFC [92]. The characterisation "reactive" or "on-demand" routing refers to the fact that the routing protocol requires participants to maintain routes only to destinations that are in active communication. Paths are established on an "as needed" basis, and there is no proactive discovery of routes. This is beneficial in mobile environments as fully up-to-date knowledge of all routes from every node implies large overhead with diminishing returns if the routes are not utilised, since topology changes may be frequent.

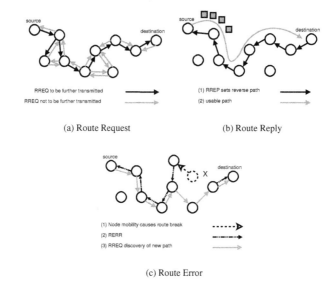

(a) Route Request   (b) Route Reply

(c) Route Error

Figure 2.1. The main operations of AODV routing

AODV makes use of a destination sequence number for each route entry. This sequence number is produced by the endpoint of a communicating pair and is included with the route information sent from the destination to requesting nodes during the route discovery process. Using this simple mechanism loop-free routing is ensured and the sequence number is further used as a metric of freshness for each route. Every time the destination notifies other nodes of a route to itself, it increases the sequence number and, hence, when the route discovery procedure presents a choice of several routes to the destination, the more up-to-date one may be chosen. In contrast to the other popular reactive routing protocol, DSR [60], AODV only maintains a single route entry per destination and in particular the freshest one as indicated by the accompanying sequence number.

There are two main functions of the routing protocol to consider, namely route discovery and route maintenance. Whenever a node is unaware of the route to its communicating destination, it initiates the route discovery process. A probing message called the Route Request (RREQ) is disseminated throughout the network via broadcasting and in a controlled fashion using an expanding ring search. The message is in turn forwarded once by each node that receives it and hence a 'forward path' is established from source to destination by imprinting next hop information onto intermediate nodes. An example of the network "flooding" with RREQ packets is shown in Figure 2.1(a) for a given topology. When the RREQ reaches the destination node or a node which is aware of a route to the destination, a Route Reply message (RREP) is sent to the node that forwarded the RREQ. When the RREP originates from an intermediate node and not the destination itself, it is referred to as a gratuitous RREP. For each RREQ cycle a single RREP is generated and in particular as a response to the first RREQ received. The RREP propagates back to the source thereby establishing a 'reverse path' where all intermediate nodes have enough next hop information to route packets for the {source,destination} communicating pair. Figure 2.1(b) depicts the process. Further, for each route entry at participating nodes there exists a list of neighbours that have made use of the route. This "precursors" list is utilised whenever route failures occur to inform only relevant nodes (i.e those making use of the route) of the route's breakage.

The second main function of AODV is its route maintenance process. After the route discovery process and so long as the discovered route is used, the routing protocol does not dictate any particular action. When the route becomes inactive, i.e. no data is sent over it, a timer is activated, after the expiration of which the route is considered stale and expires. Should the routing agent at a node become aware of a link breakage for an active route, a Route Error (RERR) packet is generated at the point of breakage. This is then disseminated to the appropriate nodes participating in the route's formation and those nodes actively using the route. The latter is achieved via the precursors list as described previously. The nodes affected by the invalid route mark it for expiration since it is no longer useful. In this fashion, the RERR message propagates to the source node which can then initiate a new route discovery phase. This operation is illustrated in Figure 2.1(c). Alternatively, the intermediate node at the point of the link failure may opt to produce a RREQ itself in expectation that the destination is still reachable. An alternate route may be discovered more quickly if the discovery process is initiated at the point of breakage rather than at the source. Such an attempt at a "local repair" is very efficient if the topology does not change radically and the destination is still reachable at a relatively short distance from the point of failure.

Finally note that provisions are also made in AODV for the discovery and circumvention of unidirectional links as well as for the use of AODV on IPv6 enabled networks.

2.2.2 Dynamic Source Routing (DSR)

DSR [60] is a distance vector routing protocol that makes use of sequence numbers to avoid routing loops. However, it is a reactive algorithm and as such does not maintain routes to all possible destinations but establishes them as the need arises. Each intermediate node does not need to contain up-to-date information for a complete path to a destination because the complete route a packet must follow to reach its destination is imprinted on its header by the source. DSR makes extensive use of route caching and as such its table entries may contain multiple routes for the same destination. Furthermore, there is no need for a mechanism to detect routing loops as loop freedom is assured by source routing.

In order for the source to discover the path to a destination the network is controllably flooded with Route Request (RREQ) packets. As a RREQ packet is rebroadcast by the intermediate nodes, the hop sequence to the destination is recorded on the packet's header. When the packet reaches the destination or a node that knows the route to the destination, a Route Reply (RREP) is transmitted back to the source by reversing the path of the RREQ packet, thus informing the source of the new route. In the case of a unidirectional link, it is necessary for the destination to initiate its own route discovery process as the inverse of the original path is not a valid path in itself. The process is illustrated in Figure 2.2.

The destination node replies to all RREQ packets received per route request cycle and, consequently, the source may discover multiple routes to the destination. Following the route discovery process, each data packet flowing from the source to the destination contains the complete hop route to the target. Whenever a link failure occurs

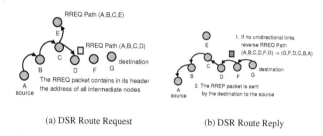

(a) DSR Route Request   (b) DSR Route Reply

Figure 2.2. Route discovery in DSR

a Route Error (RERR) packet is transmitted from the node where the link breakage occurred to the source. This is propagated through the nodes that contain the failed route in their cache which, in turn, update their caches to shorten the stored path up to the point of failure. Once the source receives the RERR packet it re-initiates the route discovery process or may make use of alternate cached routes to the destination.

It is possible to improve a node's ability to learn new routes at no additional traffic cost by allowing promiscuous listening. In this case, the node is allowed to listen to traffic not addressed to it and discover new routes from packet headers. Another use of promiscuous listening is to optimise the routing path. The listening node may examine the untraveled portion of the path in a packet header to check if its address is there. If it is, that means that the packet need not go through any other hops preceding the node's own address in the route. To make the source node aware of this, the promiscuous node then transmits a gratuitous RREP to the packet's source that includes the shortest path without those intermediate nodes. Nonetheless, intercepting all packets can be detrimental to the power reserves of the mobile node and there are computational overheads of packet processing to be considered as well [61].

The DSR protocol further includes a packet 'salvaging' mechanism. When an intermediate node forwarding a packet detects that the next hop along the route for that packet is broken but contains another route to the packet's destination in its route cache, then the node should attempt to 'salvage' the packet rather than discard it. The packet's route is altered according to the node's route cache entry for the destination and a RERR packet is sent back to the source informing it of the new route. Overall, DSR is persistent in recovering from routing errors by discovering alternate routes from the point of failure. In particular, the proposed standard [60] defines a maintenance buffer that caches packets which were unsuccessfully transmitted in view of discovering an alternate route. This mechanism can be combined with the promiscuous listening functionality as described previously so as to utilise alternate overheard (cached) routes and eventually deliver the packet to its destination.

Two other mechanisms of interest in DSR are network layer acknowledgements (ACKs) and passive ACKs. In particular, the former ensures DSRs ubiquitous nature while the latter helps reduce overhead associated with point-to-point reliable transmission of packets. The two mechanisms are now briefly described in turn.

Network layer ACKs may be demanded by the source via an option in the DSR options header embedded in a packet sent. Upon reception of such a packet, the destination responds, in turn, with a DSR ACK packet which verifies the successful data exchange. The advantage of such network layer functionality is that it enables the deployment of DSR across systems that do not feature equivalent lower layer provisions (IEEE 802.11 transceivers do, however, include such functionality [52]). Disadvantages of such a mechanism include higher ACK response time and processing overhead than in the case of MAC layer responses.

Passive ACKs occur between the two entities of a point-to-point dyad, when the packet originator overhears the neighbouring destination forwarding the packet after some time. This implies that the destination had correctly received the packet earlier, but, on the downside it requires omnidirectional antennas and both source and destination to be within transmission range of each other. This requirement holds true even when the two nodes are not conversing with each other, which in turn implies that the destination should transmit the packet using the same power as used by the source when it first transmitted the message, to ensure that the transmission will be overheard by the source. Hence, the destination node of the passive ACK pair may not opt to limit its transmission range to conserve energy when talking to other nodes [63].

It is worth nothing that the above two techniques are complementary, i.e. network layer feedback may be combined with passive ACKs. In particular, the node may at first attempt to transmit and wait for a passive ACK before resorting to the explicit request of a network layer ACK [60]. If a passive ACK is successfully received after transmission, the network layer ACK overhead is hence eliminated.

2.2.3   Optimized Link State Routing (OLSR)

The OLSR protocol [28] has been developed primarily with mobile *ad hoc* networks in mind, although no assumptions are made about the underlying link layer. Its main operation resembles table-driven, proactive protocols as used in wired infrastructure networks; in fact it is similar to the previously introduced DSDV [93] protocol with respect to there being regular topology information exchanges with other network nodes.

An OLSR network is built upon the concept of "multipoint relays" (MPRs). Each node selects a set of its neighbouring nodes as MPRs, which are then responsible for forwarding control traffic intended for distribution into the entire network. The MPRs provide a mechanism to constrain the flooded control traffic by reducing the number of transmissions required, hence mitigating DSDV's serious overhead problems [93]. Through the transmissions of its MPR 1-hop neighbours, a node is able to reach all nodes within a 2-hop radius. Nodes selected as MPRs have to feature bi-directional link status with their selector, which elegantly avoids problems associated with uni-directional communication, such as the inability to exchange link-level acknowledgements, a vital operation for 802.11 networks.

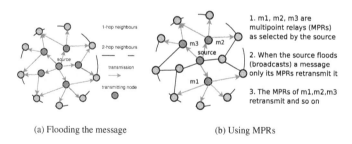

(a) Flooding the message   (b) Using MPRs

Figure 2.3. Methods of distributing control messages

MPRs are also critically responsible for announcing the link-state information of their selectors in the network. Nodes which have been selected as multipoint relays by some neighbouring node(s) announce this information periodically with topology control (TC) messages. Essentially, each node announces to the network that it has reachability to the nodes which have selected it as an MPR. When paths are calculated, the MPRs are used to form the route from a given node to any destination in the network. For MPRs to be selected, information about a node's neighbours must become available and this occurs through periodic exchanges of HELLO messages. Figure 2.3 illustrates the operation of MPRs and the resulting savings in transmissions compared to a simple flooding operation. In particular, in Figure 2.3(a) the message is sent from the source to all nodes (that is the one and two-hop neighbours) through 5 transmissions. When MPRs are employed in the same topology in Figure 2.3(b) (incurring the overhead of HELLO messages between the nodes so that MPRs may be chosen by the source) only 3 transmissions are needed for the complete dissemination of the message. It can be intuitively understood that for dense and large networks the savings are quite substantial, even after accounting for the overhead of HELLO packet exchanges.

OLSR is particularly well suited to scenarios of large and dense networks where the MPR optimisation works well, in addition to instances where the communication pairs change over time; in these cases there is no additional control traffic overhead as routes for all known destinations are maintained at all times [94]. Further, OLSR maintains sequence numbers on its control messages like AODV and DSR and periodically retransmits its control messages. This makes it as resilient as the aforementioned reactive protocols to out-of-order delivery or loss of control information.

2.3   Random waypoint model

The random waypoint mobility model [75] is one of the most popular mobility models in MANET research and in itself a focal point of much research activity [75, 83, 110]. The original version of the mobility algorithm was originally presented by Johnson et al. in [59] and refined in follow-up work [19]. The model defines a collection of nodes which are placed randomly within a confined simulation space. Then, each node selects a destination inside the simulation area and travels towards it with some speed, s. Once it has reached the destination, the node pauses for some time, p, before it chooses another destination and repeats the process. The node speed, s, of each node is specified according to a uniform distribution with $s \in (0 \dots V_{max}]$, where V_{max} is

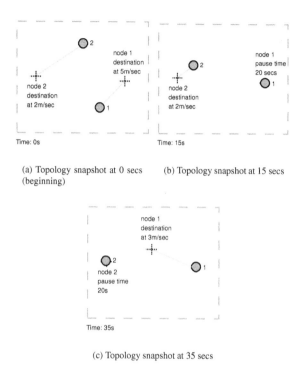

(a) Topology snapshot at 0 secs (beginning)   (b) Topology snapshot at 15 secs

(c) Topology snapshot at 35 secs

Figure 2.4. Illustration of the random waypoint mobility model

the maximum speed parameter. Pause time is a constant p secs. The cycle is illustrated in Figure 2.4. It is suggested in [19] that the simulation should be left to run for some period of time before collecting data, effectively discarding the initial observations so as to allow the probability distribution of both location and speed to converge to a "steady-state" distribution.

In the initial use of the random waypoint model for evaluation [19], an increase in mobility was simulated by increasing the maximum speed parameter or decreasing the pause time. In fact, the authors in [19] assumed that the average node speed would be $V_{max}/2$; an assumption shared by later research work [2]. However, Yoon et al. [110] have revealed that the average node speed in the random waypoint model is actually continuously decaying and further work by Navidi et al. [83] has confirmed those findings. In order to ensure that an increase in the maximum node speed parameter actually reflects an increase in average nodal speed (and by extension increased mobility) this dissertation employs the solution proposed in [110] which defines a non-zero minimum bound for the range of the uniform speed distribution and ensures quick convergence to a steady state average node speed.

It should be noted that the random waypoint mobility model is the most popular of the "entity" mobility models, where each node's motion is independent to that of others. Its popularity may be attributed to ease of implementation and intuitive appeal in view of the lack of widely deployed MANET testbeds where mobility patterns could be traced and then used in simulations. Other proposals in mobility models include several "group" mobility models [83], where the movements of nodes may be correlated, such as the motion of vehicles on the highway and so on.

2.4 Assumptions

In the subsequent chapters, extensive simulation results will be presented. The following assumptions are used throughout this research and have been widely adopted in the literature [2, 6, 24, 26, 31, 34, 35, 42, 43, 49, 65, 71, 104, 111] .

- Mobiles nodes have sufficient power supply to function throughout the simulation time. At no time does a mobile node run out of power or malfunction because of lack of power. Equivalently, the wireless transceivers are active at all times, although promiscuous listening (i.e decoding of all frames, even those not addressed to the node itself) is not active unless specifically noted.

- The number of nodes in a given topology remains constant throughout the simulation time. Note that network partitioning may still be evident during simulation and so the network may not be connected at all times. However, at no time does a node leave or gets added to the simulation area.

- Transmissions are not affected by random errors. Transmissions may still interfere with each other (i.e. affect each other if they occur in close proximity);

however a node will always successfully decode a transmission provided it is within transmission range of the source and there is no interfering transmission.

- All nodes are equipped with IEEE 802.11 transceivers. Unless otherwise stated the full RTS/CTS mechanism is employed on these wireless devices.

It is worth noting that other assumptions will be stated in the following chapters as appropriate.

2.5 Justification of method of study

In this work extensive simulations are conducted to explore issues in TCP performance in MANETs. This section discusses briefly the choice of simulation as the appropriate mode of study for the purposes of this dissertation, justifies the adoption of ns-2 as the preferred simulator and further provides information on the techniques used to minimise the possibility of simulation error.

After some consideration, simulation was chosen as the mode of study in this dissertation. Notably, when this work was undertaken, analytical models with respect to multihop MANETs were considerably coarse in nature which made them unsuitable to aid the study of TCP with a reasonable degree of accuracy; it should be noted, however, that understanding of multihop wireless communications has improved in recent times [96]. Further, since the scope of this study of TCP in MANETs involves numerous mobiles nodes, even a moderate deployment of nodes as an experimental testbed could entail substantial and prohibitive cost. As such simulation was chosen as it provides a reasonable trade-off between the accuracy of observation involved in a testbed implementation and the insight and completeness of understanding provided by analytical modelling.

In order to conduct simulations the popular ns-2 simulator has been used extensively in this work. Ns-2 was chosen primarily because it is a proven simulation tool utilised in several previous MANET studies [31, 35, 43, 49, 104, 111] as well as in other network studies [37]. While developing modifications to the simulator, special care was taken to ensure that the algorithms implemented would function as designed and that the simulator would not exhibit unwanted side-effects; this was accomplished through meticulous use of the validation suite provided with ns-2 as well as careful piecemeal testing of implemented features. Further, real-life implementations of routing agents were used for the simulations conducted in this dissertation, in order to achieve a close approximation of real system behaviour [48, 79, 84]. Finally, particular attention has been given in this work to ensure that the simulation results are fairly representative of possible real world systems [7] and to avoid shortcomings of previous research in MANETs [71].

Chapter 3

TCP performance in MANETs

3.1 Introduction

Due to its extensive use and implementation maturity on most platforms, TCP has become the focal point of much research work in MANETs [31, 35, 43, 49, 104, 111], especially with regard to the effects of non-congestion related packet losses [8, 25]. However, these efforts have largely focused on special topological configurations and the evaluation of proposed solutions has mostly been performed in limited scenarios. Furthermore, research which specifically examines the behaviour of TCP variants across different routing protocols has been meagre [2, 34] and was undertaken at a time when MANETs and their properties were not well understood. In particular, previous studies have been hampered by restrictive assumptions with regards to the mobility model used [34, 49, 95], or have been otherwise encumbered by the immaturity of available simulation tools [2, 71].

Most previous evaluation efforts have only either measured performance in static topologies [95] or through a limited number of trials insufficient to extrapolate general conclusions [2, 34, 49]. Recent work [110] has revealed an issue with the popular random waypoint mobility model, as used in the aforementioned works, concerning continuously decaying average node speed over the course of a simulation run. These considerations would seem to justify a re-evaluation of the performance of TCP agents, as previous work has not clearly reflected varying degrees of mobility [83]. Further, as the understanding of the wireless medium properties has progressed, the available simulation tools have matured and, after substantial fine-tuning, now incorporate significant detailed elements corresponding to practical implementations [7].

Motivated by the above observations, this chapter makes the following contributions. It presents a study on the effects of mobility on TCP, through its interplay with three popular routing mechanisms, namely AODV, DSR and OLSR. This provides insight into the potential performance discrepancies between routing protocols and more significantly outlines the trade-offs involved in enabling optional features included in each routing protocol. Notably, the results show that routing protocol specific features, such as packet caching or route notification interval, can significantly affect TCP performance (e.g. throughput).

This chapter also conducts a performance evaluation of three popular reactive and one proactive TCP variant, namely TCP Reno, NewReno, SACK and Vegas. Such an evaluation study is the first to examine the performance of TCP Vegas in MANETs and identify the cause of its competency. Overall, the results indicate that Vegas, SACK and NewReno maintain better goodput performance levels with respect to TCP Reno, with Vegas exhibiting the best overall goodput of the three. In contrast to previous work [49], detailed simulation traces are used to pinpoint the causes of the performance discrepancy of the TCP variants with respect to inactivity periods and reaction to packet losses.

The rest of this chapter is organised as follows. Sections 3.2, 3.3 and 3.4 include instructive simulation traces and a description of the behaviour of TCP Reno over a route break event in the presence of AODV, DSR and OLSR routing agents respectively. Section 3.5 contains the results of the performance comparison of the different TCP variants, namely Reno, NewReno Vegas and SACK, under various mobility conditions and discusses simulation parameters and assumptions. Detailed discussion of the interplay of TCP mechanisms with other layers, which accounts for the performance discrepancies observed in the simulation, are included in Section 3.6. Finally, Section 3.7 summarises this chapter.

3.2 TCP behaviour over AODV

This section examines the behaviour of TCP Reno in a low mobility scenario. This example is instructive as to the challenges faced by TCP in a wireless multi-hop environment and provides insight into the interaction of TCP with the AODV routing mechanism. In particular, the discussion that follows aids comprehension when contemplating TCP goodput performance issues in the subsequent sections. Note that due to its simplicity TCP Reno is well suited for minute simulation trace analysis and as such, used here as the transport agent. However, comments made on its behaviour are pertinent to other types of TCP agents as the discussion largely involves TCP characteristics present in all variants.

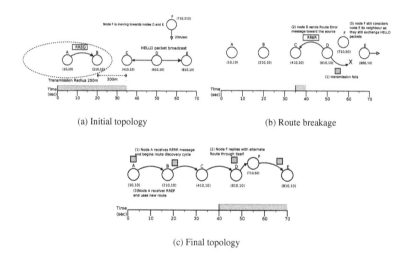

(a) Initial topology   (b) Route breakage

(c) Final topology

Figure 3.1. A scenario depicting AODV operations after a route break

Simulation setup The scenario involves a 5-node string topology (nodes $A \rightarrow E$) and an additional node, F, which stands in close proximity between nodes D and E, as shown in Figure 3.1(a). The nodes along the $A \rightarrow E$ string are spaced 200m apart and feature standard Lucent WaveLan II [64] transceivers with bandwidth of 2Mbps. An

FTP bulk transfer with an infinite backlog is initiated at the beginning of the simulation between the end points of the string topology, namely nodes A and E. As mentioned above the TCP agent carrying the FTP traffic is Reno and the segment size is 1460 bytes.

The transmission range of the transceivers is fixed at 250m using a flat, ideal signal propagation model which does not account for attenuation up to the transmission range limit and nullifies the signal strength beyond that threshold. Obviously, such a propagation model cannot occur in reality as some signal degradation is always present but it is utilised so as to isolate the effects of route breakage and disregard other effects such as those caused by interference [107]. To deal with the standard hidden terminal effect the RTS/CTS mechanism [106] is active throughout the simulation run. The routing protocol used is AODV-UU [79] which is a working AODV implementation utilised in conjunction with the ns-2 simulator [37], and which is implemented according to the corresponding RFC [92]. HELLO packets are used to detect route failures; in particular, if a node has not broadcast a HELLO packet for 5 seconds the link is considered stale. Link layer (LL) feedback is not considered in this experiment although its potential benefits and drawbacks are taken into account in the subsequent discussion. A complete list of the AODV parameters used is presented in Table 3.1. These are set as recommended in the corresponding RFCs and have been used in previously reported research [26, 34, 79, 92].

The scenario proceeds as follows. Initially, the AODV agent at node A broadcasts a Route Request (RREQ) packet which is forwarded in sequence by each intermediate node until it reaches node E, which responds by issuing a Route Reply (RREP) towards A. As the RREP propagates to the source, each node that receives it obtains enough information to forward subsequent data, as a "reverse path" is formed to complement the "forward" path set by the RREQ [92]. At the end of this phase the complete path has been setup in the form of the $A \rightarrow B \rightarrow C \rightarrow D \rightarrow E$ route. Note that node D is aware that E is its neighbour by the time the RREQ packet arrives. However, since the gratuitous RREP feature has been disabled in the AODV agents, only the intended destination may issue a reply to the discovery packet. After a few seconds and once the bulk transfer has been initiated, node F moves closer and gets stationed between nodes D and E, thus getting well established inside the transmission radii of both nodes. At the 30 second mark, node E starts moving horizontally

Table 3.1. AODV parameters

Parameter	Value	Parameter	Value
Exp. ring search	ON	TTL start	2
Local repair	OFF	TTL increment	2
Active route timeout	5 secs	LL feedback	OFF
Gratuitous RREQ	OFF	HELLO interval	1 sec

away from Node D at 10m/sec until at 35 seconds the signal of D no longer reaches E and the link becomes invalid. At its new destination node E is still a neighbour of node F but cannot be contacted by node D. Node D realises that its link to E has been invalidated when it notices the absence of HELLO packets (this is the only way to realise route breakage as MAC feedback has been disabled). A Route Error (RERR) packet is then sent back by node D to the source (node A) where a new route discovery cycle begins so as to probe for an alternate route to the destination. Note that as the RERR message propagates backwards along the route, all TCP segments buffered at each node using the invalidated route are dropped. Figure 3.1(b) outlines the repair procedure and depicts the route breakage. As soon as the RREQ packet reaches node E (through F), a RREP packet is launched from it towards the source (note once more that Node F could have generated the RREP packet but gratuitous RREPs are disabled in this scenario). At about the 39 second mark the route is restored and the TCP agent resumes transmission. There are no subsequent route breakages until the end of the simulation and the topology stands as in Figure 3.1(c).

During this delivery effort, the TCP agent is unable to distinguish among the different causes of segment loss. Certain segment losses derive from the inability of the MAC protocol to properly coordinate packet transmissions among stations, which is mostly attributed to the exponential "waiting period" backoff of the 802.11 protocol as demonstrated in [26, 104, 107]. Further, the reaction of the routing protocol to route breakages explicitly causes packet loss as each station processing the RERR packets empties its transmission queue of packets utilising the invalid route. All such packet losses are interpreted by TCP as signs of congestion despite the fact that congestion does not occur at any point during the simulation (no node's transmission queue becomes full at any time). This misdiagnosis impedes TCP performance and its effects are further compounded by reliance on the coarse grained RTO timer becoming aware of the route's restoration (there is no explicit cross-layer notification from the routing protocol concerning the route's status). Several research studies [24, 111] have shown that it is particularly desirable for TCP to maintain explicit awareness of the route's status, so that lost packets from route breakages do not spuriously activate congestion avoidance.

Figure 3.2(a) displays the DATA-ACK exchange (and thus, indirectly, the throughput) of a TCP Reno agent in the string topology scenario. Each marked ACK point in the graph corresponds to a single ACK received at the source which acknowledges a range of bytes (segments). A value of 0 denotes a duplicate ACK (since it does not acknowledge any new segments); a value of 1 denotes the normal TCP cycle since every ACK acknowledges a single additional segment (delayed ACKs are not used in this simulation). A value greater than 1 denotes that a packet which filled-in a discontinuous series of received segments at the destination's buffer was received and successfully acknowledged. The DATA segment marks at the top of the graph indicate the times when a TCP DATA segment was launched by the sender. Of particular interest is the region at 35-39 seconds (indicated by a dashed box in Figure 3.2(a)) where the ACK flow stops since the route is considered invalid. The period of disconnection, that is the period from when the routing protocol first determines that the route has became invalid until it registers its restoration, is denoted by the two solid vertical lines in the graph. Further note the discrepancy between the time of the actual route failure (at 35 seconds) and the time it takes for the routing protocol to detect it and initiate a new route discovery procedure (37.1 seconds mark). The delay is attributed to the absence of link layer feedback and the use of HELLO packets, which represents a trade-off between frequency of updates and overhead. When HELLO packets are absent for some time, the routing protocol may assume the link has been broken. However, with the link layer feedback, if a transmission fails then the route may immediately be considered obsolete without waiting several seconds for the absence of HELLO feedback to register. However, this makes the routing protocol prone to false positives from failed link layer transmissions due to interference effects or from other transient causes [25]. Apart from a stray duplicate ACK received before the RERR

notification could propagate to the source, there is no newly ACKed traffic during that
period.

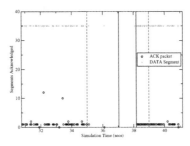

(a) Data segments sent and ACKs received

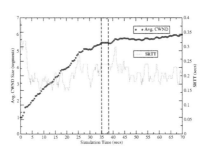

(b) Smoothed RTT and *cwnd* evolution

Figure 3.2. Goodput and RTT estimate of TCP Reno over AODV during a route break

An interesting interplay between the TCP's exponential RTO backoff can be observed in this case. The packets sent after the 35 second mark as well as ACKs in
flight are mostly lost. These losses cause TCP to retransmit at the 36 second mark after experiencing an RTO. This retransmitted packet is lost on its way at node B, which
by this time has received the RERR packet forwarded by node C. The new RTO timer
backs off exponentially and is set to approximately 2 seconds. Thus, it expires shortly
after the 38 second mark, by which time the AODV agent at node A is aware of the
route breakage and has already initiated the route discovery process. The subsequent

retransmitted TCP packet (after the RTO) is *buffered* at the source node whilst the route discovery process finds a new route. A new route is discovered 70 ms later (by the discovery process which had started earlier) and the packet is launched 30 ms afterwards by the routing agent. If a subsequent RTO had occurred, say because the route had become invalid once again, TCP's exponential back-off would have necessitated TCP to remain inactive for a longer period of time than before (approximately 4 seconds in this case) even though the route might have been repaired in the mean time. The lack of useful feedback between the routing (AODV) and the transport agents is a well known problem in MANETs and has been discussed in previous work [34, 111]. However, no previous research has mentioned that, at a rudimentary level, the negative effect of consecutive RTOs does not take place if the TCP agent launches the packet after a route breakage has been detected by the routing protocol (as in the example mentioned above). In such a case, the buffering of the packet by the routing entity circumvents the damaging effects of consecutive RTOs if the packet and its accompanying ACK are successfully transmitted once the route has been restored.

Figure 3.2(b) shows the smooth round-trip time (SRTT) measurements as realised by the Reno TCP agent. The time frame for the route failure is denoted by a dashed box in the same graph. The RTT samples freeze for some time as the route is being restored (denoted by the plateau at around 35-38 seconds in the SRTT graph). After the route has been restored, the new SRTT measurements are not significantly different to previous ones as the path is only extended by a single hop. It is noteworthy, however, that the measurements vary significantly throughout which may not be ideal for delay or jitter sensitive applications.

In the same figure, a graph of the average congestion window (*cwnd*) size is overlaid. Previous research [26, 43, 107] has revealed that TCP does not behave optimally when used under distributed MAC mechanisms such as those employed by the 802.11 protocol. Specifically, it has been shown that TCP's cwnd stabilises at a large average which maintains more packets in the pipe than is optimal [43, 107] for the MAC mechanism to function properly, especially when the path is long. In previous work [25], and for an IEEE 802.11 receiver, it has been shown that the optimal window size for a topology of 4 hops would be 1; here an average *cwnd* of 6 segments is observed.

Finally, it is of particular note that during the course of the experiments packet loss is still evident, even for packets that are not broadcast (i.e. TCP DATA and ACK segments, not just HELLO packets and RTS/CTS frames). This is surprising considering that interference is not evident in this scenario (due to the signal propagation model chosen) but the MAC mechanism nevertheless fails to coordinate transmissions effectively. Previous work has noted such losses when interference is present [32] but we confirm the phenomenon in this flat signal propagation scenario, which does not exhibit such interference.

3.3 TCP behaviour over DSR

In this section the behaviour of TCP Reno is illustrated over DSR in the same *ad hoc* scenario as the one in Section 3.2. The timing of the route breakage, the node movement and the settings of the TCP agent are identical to the previous section. The single difference in the simulation setup is that the routing agent active in every node is DSR. The particular DSR implementation used is DSR-UU [84], which is an actual testbed implementation interfacing with the ns-2 simulator. Apart from noting the reaction of TCP to the DSR routing maintenance mechanisms, this special use case describes the deployment of passive and network layer acknowledgements which have not been discussed in the literature although they are part of the DSR draft [60].

The DSR protocol, like AODV, may utilise link layer (LL) feedback to discern route failures but in this experiment such feedback is not taken into account. In particular, since LL feedback may not always be available or functioning as intended (i.e. may often result in false positives), in this simulation the DSR agent falls back to network layer and passive ACKs [60]. This combination of techniques is described in Section 2.2.2; a complete list of DSR parameters is presented in Table 3.2.

In the case studied here, the topology is initially formed as displayed in Figure 3.3(a). The route discovery procedure is initiated by the source node (node A), as outlined in Section 2.2.2 in order to find a usable path to the destination (node E). The focal point of interest is the route break that occurs at 35 seconds, as depicted in Figure 3.3(b). At that time, as in the AODV scenario in Section 3.2, node E has moved outwith the transmission range of node D, and node F is in position to act as the intermediary. The DATA-ACK packet exchanges during this time interval are shown in Figure 3.4(a) in the same format as the one used in the AODV case.

The DSR protocol functions as follows after the link break. At 35.9 seconds and

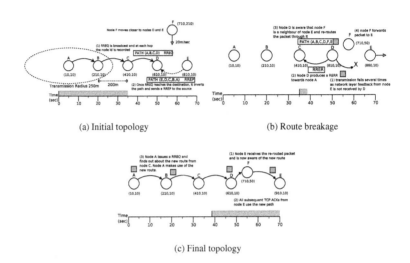

(a) Initial topology   (b) Route breakage

(c) Final topology

Figure 3.3. A scenario depicting DSR operations after a route break

Table 3.2. DSR parameters

Parameter	Value	Parameter	Value
Passive ACKs	ON	Flowstate	OFF
Promiscuous listening	ON	LL feedback	OFF
Packet salvaging	ON	Use alternate routes	ON
Snoop routes	ON	Send. Buffer Lifetime	30 secs

after several failed attempts to receive a network layer ACK, the routing agent at node D, realises that the link has been broken. Then it produces a RERR packet towards the source (node A) but, unlike AODV, does not drop packets. Instead, all the packets making use of the invalidated link are placed in the maintenance buffer and the routing agent consults its routing cache and discovers that node F is a neighbour of node E; i.e. there is an potential alternative route. The information on this route was obtained at around the first 2 seconds of the scenario when due to the inability of the MAC mechanism to coordinate transmissions, the network layer ACKs between nodes D and E failed to be transmitted and node E erroneously believed its link to node D to have become invalid. So, at node E's subsequent discovery process for a route to A, both nodes D and F responded. Node E, then opted to utilise the path proposed by node D, which offered the shortest route, but as a side-effect node D became aware of node F's neighbouring status to node E.

Subsequently, at link breakage time there is no need for node D to initiate route discovery to eventually "salvage" buffered packets that had their route invalidated by the obsolete $D \rightarrow E$ link. Instead, node D makes use of the alternate route through F by replacing the old path embedded in its packets' header with the new one. Node E also realises at 36.7 seconds that the link to D has been severed, but shortly after (at 37.2 seconds) receives the rerouted packet from node F and becomes aware of the new route ($A \rightarrow B \rightarrow C \rightarrow D \rightarrow F \rightarrow E$). This use of route caching results in some savings; in Figure 3.4(a), the area surrounded by the dashed box represents the time needed for the route to be re-established (approximately 2 seconds) which is lower than the one for AODV in the previous section (approximately 3 seconds). More importantly, the TCP agent does not experience subsequent RTOs; the route discovery and salvaging operation is fast enough for TCP to exhibit only one backoff. The caching of packets in the maintenance buffer during the route breakage as well as their subsequent forwarding also helps avoid consecutive RTOs. The effect on TCP's SRTT estimator as well as the average *cwnd* size is shown in Figure 3.4(b). As compared to AODV, the plateau in the SRTT graph (denoted by the black dashed box) shortly after the 35 seconds mark is not as noticeable, since TCP inactivity due to RTO backoffs does not last as long. Note that although in this case the use of cached routes is beneficial, the utilisation of a stale route would have had the opposite effect; the routing agent would send the packet along a non-existent path and would have to wait for a

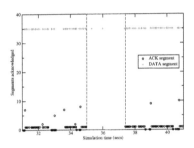

(a) Data segments sent and ACKs received

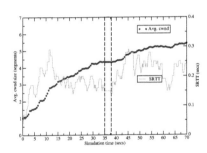

(b) Smoothed RTT and *cwnd* evolution

Figure 3.4. Goodput and RTT estimate of TCP Reno over DSR during a route break

period of time before realising that the path was invalid.

During the 35-39 seconds time frame, the RERR response which was originally produced at approx. 36 seconds by node E (after the route breakage) is propagated toward node A and causes the intermediate nodes to place packets making use of the invalidated $D \rightarrow E$ link in their respective maintenance buffers. When node D makes use of the alternate route (via node F), node C, which has already propagated the RERR on its way to the source, overhears the transmission and learns about the new route. When node A receives the RERR packet, it re-initiates the route discovery process and receives a gratuitous reply from node C. Then, subsequent packets launched from node A contain the new route and the route restoration is complete, as shown in Figure 3.3(c).

To sum up, the particular point of interest in this experiment is the avoidance of consecutive TCP RTOs due to route caching and eavesdropping. By gathering information on neighbouring nodes and routes, a usable alternate path was quickly discovered in this scenario and TCP transmissions were promptly restored. Note also, that retransmissions were avoided since packets were not dropped from the intermediate nodes but were forwarded through the alternate route when it was discovered later on.

3.4   TCP behaviour over OLSR

This section contains a description of TCP behaviour over OLSR as exhibited over the same scenario depicted in the previous two sections. All the simulation parameters are identical to the ones used previously except for the setting of the routing agent which is, in this case, OLSR. The OLSR implementation used [48] is a complete RFC [28] compliant routing daemon which interfaces with the ns-2 simulator. The particular parameters used in this scenario are depicted in Table 3.3 and are the defaults set by the reference implementation [48] and cited in the RFC [28].

As in the case of AODV and DSR, OLSR can optionally utilise link layer feedback to facilitate quick realisation of a broken link. Such feedback is disabled in this simulation run so as to prevent false positives from invalidating existing, valid routes and, thus, allow examination of TCP behaviour in isolation from such effects.

Table 3.3. OLSR parameters

Parameter	Value	Parameter	Value
HELLO interval	1 sec	TC interval	5 secs
Willingness to forward	ALL	Max. jitter	250ms
Hysteresis monitor	OFF	MPR coverage	1
Neigh. hold time	6 secs	Refresh interval	2 sec

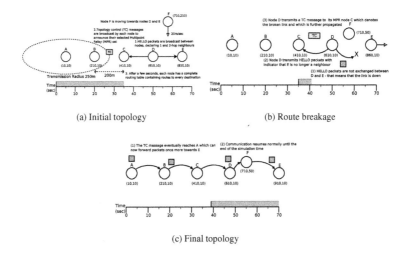

(a) Initial topology

(b) Route breakage

(c) Final topology

Figure 3.5. Scenario depicting OLSR operations after a route break

Table 3.4. MPR set chosen by each node

Node	MPR Set	Node	MPR Set	Node	MPR Set
A	B	B	C	C	B, D
D	C	E	D	F	D

Initially, and during the first few seconds of the setup, HELLO packets are broadcast from each node, which declare their immediate neighbours. After the first exchange of these messages, subsequent broadcasts also include 2-hop neighbour information. Eventually, after a short time interval, each node maintains enough information to declare a set of Multipoint Relays (MPRs) which cover its 2-hop neighbours and which is advertised using a network wide distribution of Topology Control (TC) packets. In this case, the MPR set of almost every node contains a single neighbour, as only one is necessary to reach all nodes within a two-hop radius. A notable exception is node C which has two nodes in its MPR set, namely B and D. The MPR sets are shown in Table 3.4 and the overall process is depicted in Figure 3.5(a).

Eventually, and at the 35 seconds mark, a link failure occurs between nodes D and E, due to E's movement away from its neighbour. The routing protocol's subsequent reaction is shown in Figure 3.5(b). Essentially, the absence of HELLO packets is noted in the given refresh interval (2 seconds) so the route $D \rightarrow E$ is assessed to be invalid, by the routing agents of both D and E. Therefore, packets utilising the route in D and E are dropped which leads to consecutive RTOs on the TCP agent, as the route is not restored quickly enough for incoming segments to activate the duplicate ACK heuristic. In particular, TCP experiences the first RTO at 35.6 seconds and then consecutive ones at 36.9, 39.5 and 44 seconds. The DATA-ACK exchange during that time period is shown in Figure 3.6(a), where the isolated DATA transmissions shown correspond to segment launches triggered by RTOs. Specifically, at the 35.6, 36.9 and 39.5 time marks, the DATA segments transmitted are not followed by an ACK response as they are lost upon transmission from node D to E. These particular losses are link layer losses; delivery is attempted but there is no lower level MAC-ACK response from the destination node E, as it has moved outside D's transmission radius. The unACKed DATA segment launched at about 44 seconds is discarded by node C which has yet to discover through TC exchanges the new route through node F.

The long TCP inactivity period, as denoted by the dashed box between 35-55 seconds in Figure 3.6(a), corresponds to TCP inactivity noted immediately after the route failure and occurs for three reasons. Firstly, failed transmissions are realised with the granularity of the refresh interval (2 seconds, or the equivalent of two HELLO packet launch cycles), which leads to consecutive RTO's as segments are lost in failed MAC transmissions. Secondly, the route restoration period which happens with the dissemination of TC packets has by default a coarse granularity so that several TC transmissions may be "bundled" together to avoid excessive overhead as discussed below. Thirdly, the lack of packet caching compounds the RTO issue, as ongoing TCP transmissions end up in segment drops and cause further timeouts until a new route is found. Even upon the route's restoration, TCP's RTO timer has to expire before a new "probing" segment is launched, as TCP is unaware that previous segment losses were due to link failure and attributes them to congestion.

Figure 3.6(b) denotes the smoothed RTT and *cwnd* evolution experienced by the TCP agent. The time frame of inactivity due to the route failure is denoted by the dashed box (35-55 seconds). The plateaus in the *cwnd* and RTT diagrams are noticeably larger than the ones for AODV and DSR in the previous sections, as the transfer of data stalls for a relatively longer period of time.

This long period of inactivity is due to the large TC update time interval (set to a default 5 seconds) which in turns means that information on invalid links takes several seconds to propagate. The OLSR RFC [28] makes provisions for an immediate TC packet launch mechanism which allows generation of TC packets as soon as the node's neighbourhood changes. However, since these transmissions are broadcast network-wide, there are significant overhead savings if they are "bundled" together, which makes delaying them desirable. The trade-off to consider is thus overhead against longer notification delay.

To illustrate the above point more aptly we have conducted the same experiment with reduced TC transmission interval and immediate notifications (i.e. no "bundling" of TC packets) in the same scenario. Figure 3.7 shows the total number of TCP segments acknowledged when the TC update interval changes from 5 to 3 seconds and when immediate TC updates are used. When there is a decrease in the update interval or immediate TC updates are allowed, consecutive RTO's are avoided and total time

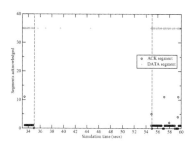

(a) Data segments sent and ACKs received

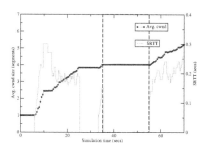

(b) Smoothed RTT and *cwnd* evolution

Figure 3.6. Goodput and RTT estimate of TCP Reno over OLSR during a route break

spent in RTO backoff is reduced as shown in Table 3.5. Note that the default update period of 5 seconds severely under-performs in this case with noticeably longer periods
of TCP inactivity (as denoted by the flat line segments of the graph in Figure 3.7).

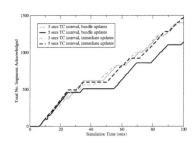

OLSR parameter	Time spent in RTO (secs)
5 secs TC interval bundle updates	55.57
5 secs TC interval immediate updates	41.07
3 secs TC interval bundle updates	41.26
3 secs TC interval immediate updates	40.38

Figure 3.7. Total TCP segments acknowledged over simulation time

Table 3.5. Time spent in RTO for different
TC intervals

Since the reference implementation [48] does not incorporate the caching of packets, unlike DSR which features a maintenance buffer, there is no forwarding of "salvaged" packets which could help avoid consecutive RTOs. Finally, the two other
plateaus in Figure 3.6(b) merit some explanation. The first one at 0-6 seconds is due to
the startup period needed for the HELLO packets and TC exchanges to take place and
setup the route. OLSR needs this warm-up period (unlike DSR and AODV which immediately start a short route discovery phase), and TCP activity does not occur until a
valid path is discovered. The latter plateau at 26-33 seconds is due to dropped HELLO
packets due to mis-coordination of the MAC protocol [107]. This phenomenon has
been observed in the other two routing agents but the slower route restoration and lack
of packet caching of OLSR causes TCP to under-utilise the route for longer in such
occurrences.

Eventually, after the TCP RTO timer expires, transmission is resumed. At simulation's end the scenario topology has settled in the form presented in Figure 3.5(c).

To sum up, the effects of consecutive RTOs are more pronounced in the case of
OLSR since, by default, it features a conservative route restoration mechanism, which
is tuned for dense networks servicing multiple connections. Shortening the refresh

interval (by tuning the TC parameter) in this case can improve performance significantly, although there is a trade-off of extra overhead against improved throughput performance to consider.

3.5 Performance evaluation of TCP variants

This section contains the results of our evaluation of Reno, NewReno, SACK and Vegas in dynamic MANET topologies. First, the simulation parameters are presented and discussed in depth. Then, the simulation results on the performance of these TCP variants over different routing protocols are listed and examined in turn.

3.5.1 Simulation setup

The general evaluation of the various TCP agents in this section is conducted with the ns-2 simulator [37]. Although the simulation setup is verbosely described below, a complete list of the routing agents configuration as well as an outline of TCP parameters are included in Appendix A.

Simulation area and mobility model: Simulations take place over two types of flat areas; a flat square arena with dimensions set to 1000x1000m and a flat strip area set to 1500x300m. In both cases, 50 nodes are placed randomly in the arena. This setup mirrors previously reported research [19, 26, 34, 49, 104]. The mobility model used is the random waypoint model [75] with parameters set to reflect mobility ranging from walking (approximately 2 m/s) to vehicular speeds (approximately 20 m/s). The simulation parameters are portrayed in Table 3.6. Note that an increase in the setting of maximum speed in the random waypoint mobility model is not necessarily indicative of a significant increase in the mean node speed as previously believed [49]. Research by J. Yoon et al. in [110] contains a thorough discussion of the issue and proposes a solution which is applied to the topologies used in these simulations. It is worth noting that previous TCP evaluations over MANETs [26, 34, 49] have not considered this limitation of the random waypoint model. For clarity, in these simulations, the mean node speed for the topologies used are shown in Table 3.6 inside parenthesis next to the maximum node speed parameter.

TCP transfer setup and metric used: For each simulation run a TCP connection is set between two randomly selected nodes to facilitate an FTP transfer session for the duration of the simulation. Hence, there is a single source of TCP regulated traffic in the network. The TCP segment size and other parameters are set as in Table 3.6. The performance metric is *goodput* which is defined as the number of packets successfully transmitted by the sender for which an ACK has been received. Retransmissions (spurious or otherwise) do not contribute to the metric; each segment's contribution to the advancement of TCP's "sliding window" is only measured once. For each pause time and maximum node speed combination the average goodput of 50 topologies is calculated and a 90% confidence interval for each is produced (shown as standard error bars depicting standard deviation in the relevant figures). The different TCP variants are analysed on the same topologies, and the same source/destination pairs are chosen per trial so as to ensure fairness and relevance of the results. A paired t-test is performed on the observations to determine if there are statistically significant differences in the performance of the TCP agents. The t-test used in this case is Welch's t-test [44, 101], which assumes that the means (but not the variances) of the normally distributed populations are equal; however, for the topologies used in this study, the results and subsequent conclusions also hold if the assumption of normality is dropped and the non-parametric Wilcoxon signed-rank test is used.

As in previous studies [19, 34, 49], the overall simulation (and connection) time is set to 900 seconds.

The signal propagation model used is the Two-Ray Ground model where signals propagate from sender to receiver in an open environment and over two possible paths; one by a direct ray and one that is reflected from the ground [102]. Essentially, this model is representative of environments where a strong line of sight is present but ground reflections also influence path loss. This is the standard propagation model used in TCP evaluation over MANETs [43, 65, 106, 107], although there are several others such as Shadowing and Ricean/Rayleigh fading [62, 99].

For the sake of clarity, a few of the simulation parameters merit some discussion. The simulation time, set to 15 mins, is chosen in order to examine TCP performance over bulk file transfers also reflected by the choice of the traffic source (FTP with unlimited backlog). It is an open research question whether TCP variants in MANETs

Table 3.6. Simulation parameters

Parameter	Value
Pause Times	0 secs (continuous mobility)
Max. Node Speed	2, 5, 10, 15, 20 m/s
Mean Node Speed	2(1.44), 5(3.34), 10(6.62) m/sec 15(9.10), 20(13.78) m/sec
Simulation Time	900 secs
TCP parameter	**Value**
Min. RTO	200ms
Max. RTO	60secs (RFC 2988)
RTO Timer Granularity	10ms (Linux kernel 2.4)
Maximum burst per ACK received	3 segments
Delayed ACKs	disabled
Segment size	1460 bytes

perform differently under other types of traffic load. However, the focus of this dissertation is the behaviour of different variants when the full spectrum of their congestion avoidance mechanisms is utilised over substantial time periods. Transferring small files such as web pages may only activate the slow start mechanism which is identical in all the examined variants apart from Vegas. Further, note that the simulation time and traffic patterns chosen are in harmony with previous related work [25, 34, 49, 87].

The choice of granularity of the RTO timer as well as the minimum and maximum RTO settings are also noteworthy since previous work has shown the detrimental effect of the RTO's exponential backoff mechanism in MANETs when consecutive losses occur due to spatial contention [43, 107]. A proposed solution offered in [34] has been to freeze the RTO timer after 3 consecutive backoffs, but such an action raises concerns of congestion if widely adopted. The RTO parameters chosen for the simulation in this chapter are representative of modern operating systems and conform with the relevant RFC [89]. Further, the choice of setting a maximum burst parameter, as shown in Table 3.6 defines the maximum number of packet launches an ACK may trigger upon reception. This helps reduce "burstiness" in TCP behaviour across the different variants, as has been discussed in detail in [36] and widely deployed in practice [4].

Finally, the choice of the particular TCP agents under examination may be justified as follows; TCP Reno is the earliest TCP variant which implements the fast retransmit/fast recovery mechanism adopted by all its modern derivatives and, as such, is used as the baseline for comparison. TCP NewReno and SACK are more modern Reno derivatives and are widely deployed, with the latter requiring both receiver and sender side modifications of the standard Reno mechanisms. TCP Vegas has been shown to have desirable properties for MANETs [108] but prior to this work has not been evaluated on dynamic topologies. Finally, note that recent research into the types of TCP agents deployed on the Internet [4] confirms the TCP parameters used in this study as widely deployed defaults in operating systems and, thus, representative of a typical client.

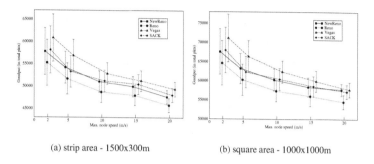

(a) strip area - 1500x300m   (b) square area - 1000x1000m

Figure 3.8. Goodput against maximum node speed for different TCP agents over AODV

3.5.2 Performance results and discussion

The goodput results for the TCP agents under examination over the AODV, DSR and OLSR protocols, namely Reno, NewReno, SACK and Vegas, are depicted in corresponding graphs as a function of goodput over maximum node speed.

TCP performance over AODV

The graph in Figure 3.8(a) indicates the superior goodput of TCP Vegas as evident against the reactive variants in strip (1500x300m) topologies. This performance discrepancy is significant and ranges from 5-9% against TCP Reno, while it narrows against NewReno and SACK (to 2-5% and 2-6% respectively). However, the difference in goodput between Vegas and the other TCP agents remains significant particularly at low mobility conditions, i.e. for maximum speeds of 2 and 5m/sec, where it ranges between 5-6%. Note, however, that the performance gap could potentially diminish if the TCP agent is adversely affected by other factors, such as spatial contention caused by other traffic in the network. Nonetheless, the results presented here confirm the competent performance of Vegas under various mobility conditions when such interaction is not considered. This expands on previous work on static topologies [108] and reveals that Vegas' performance merits are equivalent or more pronounced (in the case of low mobility) to NewReno and SACK's when single connections are considered. It is further noteworthy that TCP Reno is a significantly worse performer compared to all other variants; an observation which holds true under all mobility conditions. NewReno and SACK perform comparably in terms of goodput as their respective difference in performance is at most $\sim 1\%$ (at 5m/s). Finally, it can be observed that for each TCP variant as link breakages become more frequent due to increased mobility, the achieved goodput is reduced. This is denoted by the declining trend of the graph in Figure 3.8(a) as the mean node speed increases. This observation corroborates previously reported research [34, 49], and is also evident for the other routing protocols, i.e. DSR and OLSR, as will be revealed below.

The goodput results for the square topologies are shown in Figure 3.8(b). The performance merits of TCP Vegas are still pronounced at lower mobility conditions (max. speeds of 2 and 5m/sec), as it improves $\sim 5\%$ upon NewReno/SACK and $\sim 10\%$ over TCP Reno. This difference becomes smaller at 10m/sec (to approx. $\sim 3\%$) and diminishes at very high mobility (15 and 20m/s) where there is no significant difference between the top three variants (SACK, Vegas and NewReno). Consistently, TCP Reno under performs with respect to the other variants in the range of 5-10%.

TCP performance over DSR

The simulation results for DSR, closely mirror those of AODV in the previous subsection. In Figure 3.9(a) and 3.9(b) the goodput results for Reno, NewReno, SACK and Vegas are portrayed for simulation runs on a strip area. Vegas exhibits a substantial 7-12% goodput improvement over TCP Reno and $\sim 5\%$ over SACK/NewReno across various mobility conditions. Conversely, TCP SACK and NewReno have statistically insignificant difference in performance, for all node speeds, whilst Reno achieves consistently the worst goodput of all variants; a disadvantage that ranges between 5 and 12% compared to the rest.

In square topologies, the goodput trend of the TCP variants ranges as shown in Figure 3.9(b). TCP Vegas maintains a 6-10% lead over Reno, 2-5% over NewReno and 2-6% over SACK. SACK and NewReno are again comparable to within $\sim 1\%$

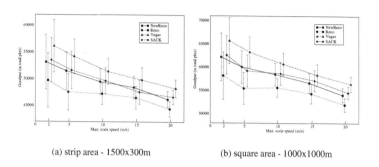

(a) strip area - 1500x300m   (b) square area - 1000x1000m

Figure 3.9. Goodput against maximum node speed for different TCP agents over DSR

worth of performance discrepancy. Reno is significantly the worse performer incurring a 4-10% performance penalty over the other variants at various mobility conditions.

TCP performance over OLSR

Figure 3.10(a) shows the goodput performance of the TCP variants over OLSR in strip topologies. TCP Vegas, as in the case of AODV and DSR, maintains a performance advantage over Reno at a 7-9% margin and over SACK/NewReno at 4-6% for low to medium mobility conditions (2, 5 and 10 m/sec). SACK and NewReno, perform similarly, within a $\sim 1\%$ difference margin of each other. TCP Reno is the worst performer, exhibiting a disadvantage of 3-9% throughout different node speeds over the other variants. A point worthy of note in this case is the closing of the performance gap between Vegas and the rest of the variants under high mobility (15m/s) which is $\sim 2\%$.

Figure 3.10(b) outlines the goodput results in square topologies. TCP Vegas exhibits again the highest goodput, with a 7-10% goodput improvement over Reno and 4-8% over SACK/NewReno. NewReno and SACK do not exhibit differences in performance of over $\sim 1\%$. Finally, Reno achieves the lowest goodput by under-performing 4-10% compared to the other variants and for all mobility conditions.

The results, as presented above, reveal a trend across routing protocols where TCP Vegas has superior goodput over the reactive variants especially under low mobility conditions. Previous research [43, 107] has suggested that TCP agents suffer from

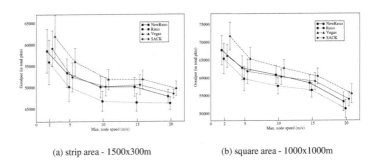

(a) strip area - 1500x300m   (b) square area - 1000x1000m

Figure 3.10. Goodput against maximum node speed for different TCP agents over OLSR

throughput instability in *ad hoc* networks as packet loss are caused by the MAC mechanism's inability to properly utilise the shared medium. During the experiments above it was noted that all TCP agents suffered from frequent retransmission timeouts (RTOs), even at low mobility speeds. The next section discusses the impact of RTO's on each variants' performance and provides some insight on the performance merits of TCP Vegas.

3.6   Discussion on TCP mechanisms

This segment aims to describe in detail the extensive retransmission timeout (RTO) phenomenon which decreases overall TCP goodput for all the variants examined in the section above. Subsequent conclusions are drawn after considering simulation traces in the scenario of a static topology.

To facilitate discussion, a setup of a string topology of 5 hops is assumed as a special case of a long path. An FTP transfer is initiated between the end points at the start of the simulation and ceases at the 140 seconds mark. The routing agent used is AODV and the rest of the simulation parameters are as in Section 3.2. Even though the following discussion is limited to AODV, the principal observations are applicable to other routing protocols featuring similar packet caching or "salvaging" paradigms.

Figure 3.11(a) shows the 100-moving (or running) average of the number of segments in flight during the duration of the FTP session. Note that the first and last 50 observations are not averaged, since an n-moving average of a sequence of N elements contains $N - n + 1$ elements; i.e. in this case the initial and trailing 50 observations are not smoothed. Further, note that the graph in Figure 3.11(a) indicates the *actual* number of TCP data segments in flight as opposed to the *cwnd* value at each agent (i.e. how many segments the agent *estimates* as being in transit [5]). As depicted, the number of segments in flight maintained by Vegas are noticeably and significantly fewer than those of the other agents, i.e. Reno, NewReno and SACK.

Figure 3.11(b) displays the discrepancy in the *cwnd* estimate and the actual segments in flight of the NewReno and Vegas agents. It is, hence, apparent that the NewReno agent mostly *underestimates* the number of segments in flight as opposed to the Vegas agent which mostly *overestimates* them. The trend is similar for Reno and SACK, both of which mostly overestimate the segments in flight, but which are not portrayed in this figure for clarity. This overestimation occurs because the routing protocol "salvages" packets when there is indication of route failure (in the case of AODV, the local repair mechanism has to be enabled in order for "salvaging" to occur). In such an instance, Reno-based TCP agents, which are more aggressive in their transmission rate than Vegas [108], often retransmit segments that are already in flight, after the RTO timer expires. This, in turn causes, more spatial contention, which results in an increase in the number of RTOs, leading to lower goodput.

To validate the above observation we consider the number of RTOs experienced by each variant in this scenario, as well as the time spent in inactivity (RTO backoff period) as shown in Table 3.7. On the same table a measurement of the goodput achieved by each variant is shown as a total number of segments delivered as well as the percentage of goodput improvement of each variant over the baseline performance of TCP Reno. It can be deduced that as in the case of the general topology scenarios examined in Section 3.5.2, the Reno agent spends substantially more time in RTO than Vegas, NewReno or SACK. Moreover, even though TCP NewReno maintains more segments in the pipe than Reno (as shown in Figure 3.6), its ability to recover from multiple losses within the same congestion window allows it to avoid multiple or consecutive RTOs. Hence, even though NewReno creates greater spatial contention, it nonetheless experiences less inactivity time, than Reno and even SACK.

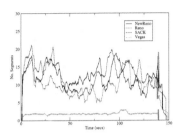 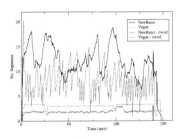

(a) Moving average of DATA segments
in flight maintained by each TCP agent

(b) Estimated (*cwnd*) vs actual segments
in flight for NewReno and Vegas

Figure 3.11. Simulation traces of a single TCP connection in a 5-hop string topology

TCP Vegas, overall, exhibits the highest goodput as it experiences only 7 RTOs and incurs the least idle time of all variants. TCP SACK which makes use of more extensive feedback than the other variants, is still outperformed by Vegas and performs on par with NewReno. Notably, the explicit knowledge of TCP SACK of which "gaps" in the destination's receiving buffer are to be filled does not result in noticeable improvement compared to NewReno. Closer inspection reveals that even though the SACK advice is utilised 138 times in this scenario, the average number of segments in flight and hence "spatial" contention are not significantly affected and are not substantially different from NewReno as shown in Figure 3.11(a). However, even though SACK experiences far more RTOs than NewReno (19 vs 9) and spends more time being idle (18 vs 11 seconds), it does not perform substantially worse; overall its retransmission strategy is not noticeably better or worse than that of NewReno. It is worth considering that the duplicate ACK responses of a SACK TCP receiver, are larger in size than those of the other variants due to the to the extra space required to accommodate the SACK-block information. As ACK segments compete with DATA segments for transmission time, the larger ACK size leads to greater spatial contention, however, this does not lead to significantly worse performance than NewReno.

Finally, consider that in the case examined here the minimum RTO is set to 200ms; actual implementations may have this parameter set to 1 sec, adhering to the original RFC [54], which would aggravate the impact of RTOs on goodput and make the

Table 3.7. RTO inactivity and goodput for TCP agents

TCP Agent	Time spent in RTO	Number of RTOs	Total Goodput (in pkts)	Perc. increase comp. to Reno
Reno	29.85	25	2253	N/A
NewReno	11.33	9	2484	10%
SACK	18.4	19	2431	8%
Vegas	10.3	7	3290	22%

performance gap between Vegas and the reactive variants even more evident.

To conclude, TCP Vegas has been explicitly shown to experience little inactivity time compared to the other TCP variants in this scenario which along with the observations made in Section 3.5.2, lead to this being identified as a source of its competent goodput performance. NewReno and SACK perform equivalently to each other, even though SACK expends more time being inactive. All variants are substantially better that Reno which is penalised severely from sending inactivity caused by RTOs. These effects can largely account for the performance differences presented in Section 3.5.2.

3.7 Conclusions

This chapter has examined the performance of a proactive (Vegas) and three reactive (Reno, NewReno, SACK) TCP variants over dynamic topologies in MANETs under three popular routing protocols, namely AODV, DSR and OLSR. The discussion has included detailed simulation traces of a route breakage scenario where it has been shown that the adverse effects of RTO back-offs may be mitigated through caching by the routing protocols. In particular, it has been shown that the AODV caching of packets at the source during the route discovery phase helps avoid multiple RTO backoffs when a route is being rebuilt. In the case of DSR, the caching of packets at intermediate nodes has been demonstrated to be beneficial if an alternate route is quickly established by the routing cache mechanism of an intermediate node after the route break. OLSR has been shown to cause significant inactivity periods for the TCP agent as its route update mechanism is, by default oriented for use over dense *ad hoc* networks characterised by multiple simultaneous flows. A solution was offered to this problem by reducing the time interval between route updates or disabling the "bundling" feature for such updates.

TCP Reno, NewReno, Vegas and SACK have been evaluated in dynamic topologies over square and strip simulation areas. TCP Vegas has been shown to outperform Reno in terms of goodput by margins of 5-12% in low, medium and high mobility conditions under the AODV, DSR and OLSR routing protocols. Vegas also exhibits a smaller, but still noticeable performance advantage over TCP NewReno and SACK in the range of 6% in low mobility conditions and approximately 2-4% performance advantage in medium and high mobility environments.

Moreover, all TCP agents have been shown to suffer from multiple RTO backoffs, but TCP Vegas and NewReno have been noted to be affected the least in a string topology of moderate length. In this particular case, TCP Vegas exhibited the fewest RTO backoffs, whilst the better segment loss handling ability of NewReno was shown to be the decisive factor of its goodput superiority against Reno. Specifically, although NewReno maintained more segments in-flight than Reno, it nonetheless experienced fewer RTOs because of its ability to recover from multiple losses within a single window of data. TCP SACK was shown to deliver comparable performance to NewReno, even though it experienced more idle time and more RTOs; the equivalence in performance merit was attributed to the former's better packet loss handling mechanism through the use of the extended SACK information.

As the competent goodput performance of TCP Vegas in the experiments was driven by its conservative congestion avoidance mechanism which maintains fewer packets along the path than the reactive TCP variants, an enquiry may be made into whether the mechanism can be adapted into the reactive variants to yield similar benefits. Such an enquiry is of significant interest as reactive (Reno-based) TCP variants are less computationally intensive than proactive ones (such as Vegas) and more widely deployed. To this end, the next chapter proposes a Vegas-inspired mechanism applied to Reno-based variants which follows a more conservative policy of packet injection into the network. The implications of such a mechanism include a reduction of packet loss due to hidden terminals and an increase in TCP goodput. The proposed changes are further thoroughly discussed with respect to ease of deployment and are evaluated in a variety of mobility conditions.

Chapter 4

TCP and spatial reuse in MANETs

4.1 Introduction

In MANETs, access to the shared medium is coordinated with a distributed MAC mechanism [52], which includes provisions to avoid the *hidden terminal* effect. This occurs when two stations do not manage to coordinate their transmissions such that they overlap time-wise to some degree. The result is a collision as the superposition of the signals becomes meaningless and transmission bandwidth is wasted. The IEEE 802.11 protocol offsets this issue by using a virtual carrier sense indicator, which is set by short request to send/clear to send (RTS/CTS) frame exchange between communicating nodes [52]. However, these provisions are not always effective in practice and could become counterproductive [103]. In particular, the discrepancy between a node's transmission (range within which other nodes can properly decode a transmitted signal) and interference ranges (range within which the signal cannot properly be decoded but may interfere with other signals) produces hidden terminals even if the RTS/CTS mechanism is in use [103]. As noted in [107], this phenomenon can decrease TCP throughput severely for the sender even when there is only one connection present in the network and can further interfere with the operation of the underlying routing protocol [106]. When segments are dropped because of such effects and not buffer overflow, as is the case in wired networks, the loss is attributed to *spatial contention* [26, 43, 103, 107].

Earlier research by Xu et al. [106, 107] has demonstrated that when TCP commands

71

back-to-back transmission of segments across a long enough path, hidden terminal effects become evident as segments are distributed along the path in a pipelined fashion. The authors in [107] have suggested that the degradation in TCP throughput caused by the spatial contention aggravated by these effects can be dealt with by limiting the TCP congestion window (*cwnd*) to approximately four segments in their considered scenarios. Follow up work by Kanth et al. [65] has confirmed these findings and suggested altering the 802.11 MAC backoff mechanism to give competing nodes a greater window of opportunity to gain access to the medium. Fu et al. [43] have also studied the phenomenon and produced an approximate estimate of the optimal TCP *cwnd* for string, cross and mesh topologies. The same study has suggested two link layer schemes to improve performance. The results in [43] and [26] have suggested that the optimal use of the wireless medium is dependent on the ability of nodes to transmit simultaneously as long as they are outside each other's interference range. Overall, as nodal coordination is not particularly effective in 802.11 multihop networks, the above cited work have demonstrated how the transport agent may be adjusted in such a way as to "coerce" the MAC protocol into more efficient operation.

The previous chapter has revealed the relative merits of TCP Vegas over Reno-based variants due to its conservative *cwnd* evolution, which maintains fewer segments-in-flight and avoids the detrimental (throughput-wise) effect of consecutive RTOs. Motivated by this observation, this chapter proposes reducing the sending rate of Reno-derived TCP variants during the slow start and congestion avoidance phases as an effective approach for dealing with the degrading effect of hidden terminals due to interference in MANETs. Such an approach mimics the conservative *cwnd* evolution of TCP Vegas, by adjusting a few parameters in standard Reno-based agents, and leads to higher goodput performance especially along lengthy paths, as will be shown below. The main motivation in choosing to improve Reno-based variants rather than explore Vegas further, is the popularity of the former [4] as well as its lower computational requirements [16] which, in turn, imply lower power demands on the possibly limited power reserves of the MANET nodes.

The contributions of this chapter are three-fold; first, an introduction is offered on the interaction of TCP agents and the routing protocol in the presence of hidden terminals, which provides insight and summarises conclusions of previous research on the subject [26, 43, 65, 107, 108]. Then, a study on emulating Vegas behaviour in standard

Reno-based TCP variants is conducted, by adjusting TCP's behaviour during the slow start and congestion avoidance phases. This results in determining an effective combination of parameters leading to the Slow Congestion Avoidance (SCA) TCP variant. The new TCP agent is then compared to an existing solution to mitigate the effects of hidden terminals, proposed in [26], which suggests limiting the maximum *cwnd* value. This comparison is performed by applying both techniques in scenarios featuring dynamic mobility patterns and then interpreting the results. Finally, considerations on applying the technique to multiple TCP flows and utilising routing layer feedback are addressed.

The rest of this chapter is organised as follows. Section 4.2 presents an overview of TCP behaviour with respect to the hidden terminal effect and its contribution to spatial contention. Section 4.3 presents an enquiry into ways of introducing a more conservative transmission rate increase in Reno-based TCP variants during the slow start and congestion avoidance phases. This investigation leads to an effective combination of "slowdown" mechanisms which is termed SCA TCP and used in the subsequent performance evaluation. Section 4.4 includes the results of a performance evaluation of the newly introduced conservative TCP variant in dynamic topologies and further entails a comparison against a popular existing solution to alleviating spatial contention, namely the adaptive Congestion Window Limit (CWL) method. Then, Section 4.5 discusses the application of SCA to multiple TCP flows and includes considerations on incorporating feedback from the routing protocol to adjust the SCA slowdown parameter. Finally, Section 4.6 summarises the results of this chapter.

4.2   TCP and spatial reuse

This section includes a brief description on the interaction of TCP agents with spurious segment losses as caused by spatial contention due to hidden terminals. Subsequent discussion is distilled through an example on a string topology as commonly done in several previous research works [26, 43, 103, 107].

For the purpose of the present discussion, consider five nodes arranged in a static string topology fitted with identical wireless devices and with the distance between any two successive nodes set to 200m, as shown in Figure 4.1. Each node can communicate with any neighbours inside its communication (transmission) range as indicated by the dotted lines. Moreover, each node exhibits a certain interference range which is the distance that its signal can be detected as a transmission but cannot be decoded properly [64]. The interference range depends on the sensitivity of the receiver as well as the wave signal propagation model used. In this example, the Two-Ray Ground signal propagation model is considered [102]. In this case, the interference range for each transmitter extends to 550m and is indicated in Figure 4.1 by a dashed line for node D. The transmission range of each wireless device is 250m and is indicated by dotted lines in the same figure. These device characteristics model the standard Lucent WaveLan II wireless transmitters [64] as used in previous research studies [26, 43, 65, 106, 107].

For the string topology, there exists only one path from any node to any destination, which includes all intermediate nodes. Hence, if node A were to communicate with node E, the transmission path $A \rightarrow E$ would include every node in between (namely, nodes B, C and D). As segments travel between nodes, each segment transmission interferes with the transmissions by other nodes inside a 2-hop radius around the transmitter. The RTS/CTS exchange, as defined in the 802.11 standard [52], can only inform of impending transmission the nodes that are inside a 1-hop radius around the communicating parties, i.e. can only provide information to nodes inside the transmission range of the sender and the receiver. Hence, nodes outside the transmission range but inside the interference range of the sender or the receiver do not receive any information about pending transmissions which can cause them to transmit at the same time and cause a collision. Furthermore, nodes inside the interference range of transmitters cannot correctly decode segments that originate from other senders (who are unaware that another conversation is taking place and their signal is not being received by the intended destination).

Consider the case of an $A \rightarrow E$ transmission route depicted in Figure 4.1. As segments travel along the path, interference causes several segment drops, as shown in previous work [65, 107]. For instance, if node D was transmitting to node E, node A would be unaware of the transmission and would attempt to transmit to B, even though B, which would be inside node D's interference range, could not obtain the transmitted frame. This is a typical case of the hidden terminal effect where node D is the hidden terminal with respect to A. Even if the frame was successfully received, it would be doubtful that B would be able to transmit a MAC layer ACK back to node A

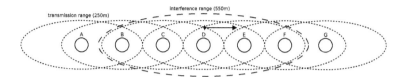

Figure 4.1. Node D transmits to node E. Node A may transmit at the same time to B even if node D has performed an RTS/CTS exchange, because A is outside the transmission range of D (but B is inside the interference range of D). Node D is "hidden" with respect to node A.

because it would detect D's communication (but be unable to decode it properly) and would defer transmission. As node D is part of the $A \rightarrow E$ path, it is likely that it would subsequently attempt to forward several segments along the path to E, and thus compound the problem. The issue of interference under discussion in this case, would be the result of a single end-to-end conversation ($A \rightarrow E$) and in the case of the TCP protocol would be further aggravated by the delivery of ACKs from the destination to the source which also contend for access to the shared medium [107].

Note that it is not necessary for the interference range to be more than twice the size of the transmission range (as in the previous example) in order for interference effects to appear. It is enough, when a transmitting source/destination pairing is considered, for the destination to be inside the interference range of another node which is outside the destination's transmission range (so it cannot decode the CTS frame transmitted by the destination to the source). Possible transmissions from that "interfering" node result in signal conflicts at the destination which cannot properly decode segments sent from the source [63].

Fu et al. [43] have noted that segment collisions due to interference can be avoided if transmissions are coordinated in the path string of nodes in such a way so that the transmitting nodes are always outside each other's interference range. Successful transmission coordination is referred to as *spatial reuse* in the same work and it is desirable to maximise this property (i.e. perform as many simultaneous non-interfering transmissions as possible) in order to improve throughput. In the example used here, the maximum spatial reuse is achieved at 1/4 of the string length, i.e. simultaneous transmissions can occur if the transmitting nodes are 4 hops apart (eg. $A \rightarrow B$ and $E \rightarrow F$). The spatial reuse factor of a given path depends on the interference range which in turn is related to the propagation model used for the path loss of the signal. Xu et al. have demonstrated in [103] the universality of the interference issue by demonstrating that in an open space environment the RTS/CTS exchange becomes ineffective due to interference as the distance between transmitter and receiver exceeds $0.56 * R_{tx}$, where R_{tx} is the transmission range.

An important implication of the MAC layer frame drops caused by interference is the rogue feedback provided to the routing protocol. As noted in [107], the routing implementation is allowed to make use of link layer feedback to detect broken routes, and actual implementations do so [79]. In particular, if seven consecutive RTS/CTS (or four DATA) transmissions fail [52], the segment is dropped from the interface transmission queue of the sender and the routing protocol is allowed to interpret that event as a sign of route breakage. Normally, this enables the sender to realise that the link has been broken much quicker as opposed to noting the absence of "HELLO" packets from the receiver, which could take several seconds[1]. However, as the RTS/CTS (or DATA) frame drops may have been caused by hidden terminal interference, the route may not have truly become obsolete and route discovery need not be re-initiated as the original route can still be used. The case of the string topology in Figure 4.1 has been examined in the literature and it has been shown to lead to several spurious route breakages [107].

Overall, the hidden terminal effects caused by interference severely affect TCP performance leading to consecutive RTOs and underutilisation of the medium. Chapter 3 in this dissertation included detailed simulation traces of the phenomenon in string topologies and outlines the effects of segment drop on TCP for three popular routing protocols (AODV [92], DSR [60] and OLSR [28]).

4.3 Proposed modifications to TCP

Chapter 3 has revealed the relative performance merits of TCP Vegas with respect to Reno-based variants (Reno, NewReno and SACK). In the same chapter, it has been shown that Vegas' conservative *cwnd* increase allows it to make more optimal use

---

[1]HELLO packets are widely used in reactive routing protocols, and so this observation holds true for AODV [92] and DSR [60]. In practice, even proactive protocols use some type of infrequent handshake to ensure determination of link validity [28].

Table 4.1. TCP parameters

TCP Parameter	Value
Min. RTO	200ms
Max. RTO	60secs (RFC 2988)
RTO Timer Granularity	10ms (Linux kernel 2.4)
Maximum burst per ACK received	3 segments
Delayed ACKs	disabled
Segment size	1460 bytes

of the wireless medium (by maximising spatial reuse), which corroborates previous related research [45, 72]. This section reveals the results of an enquiry into introducing a more conservative *cwnd* increase into Reno-based variants without compromising their reactive nature or congestion avoidance efficiency. Subsequent sections 4.3.1, 4.3.2 and 4.3.3 examine the application of more conservative sending rate increase paradigms in the slow start phase, congestion avoidance phase or both in Reno-based TCP variants, respectively. At the end of the enquiry, the most suitable of the phases is chosen as an effective place to introduce a slowdown mechanism.

Simulation setup: The subsequent examination entails string topologies set up as in Section 4.2 and as used in the previous chapter. The signal propagation model and transceiver characteristics are also the same as outlined in Chapter 3. Detailed parameters with respect to the routing protocol used are included in Appendix A.1. Common TCP parameters are outlined in Table 4.1.

4.3.1   Slow start modification (SS TCP)

The slow start phase in Reno-based TCP variants commands an *exponential* increase in the congestion window (*cwnd*) size. Specifically, for every ACK received that acknowledges new data, *cwnd* may be incremented by at most the number of bytes in a full sized segment[2] [5].

In standard TCP implementations the above directive leads to an increment in *cwnd* by the maximum allowed amount of bytes [11, 76]. In order to "emulate" a more conservative increase in the sending rate during that phase, it is possible to define a smaller increase while still retaining RFC compliance. Mimicking the approach of TCP Vegas in this regard, where the sending rate increase during slow start occurs every other ACK received [16, 17], we similarly define a delayed increase.

Specifically, our modifications are shown in Algorithm 1. The variable *s_increase_thresh* sets the number of ACKs that need be received before *cwnd* increases by a full sized segment. For instance, a value of one for *s_increase_thresh* precisely emulates the slow start behaviour of TCP Vegas by increasing the sending rate every other ACK received. Note that the new method takes advantage of the self-clocking property of TCP and does not introduce any overhead in the form of extra timer requirements. In actual implementations, this would translate to little overhead being introduced, which in turn would add little to power and computational requirements.

To evaluate the effectiveness of the proposed changes, we have conducted experiments on string topologies. An FTP connection with infinite backlog is initiated between the end-points at the beginning of the simulation and lasts throughout. Two TCP agents are evaluated, namely TCP Reno, which exemplifies a base case scenario and TCP NewReno. The overall simulation time is set to 900 seconds and the metric collected at the end of the simulation is the average goodput (in bits per sec) achieved by TCP. The definition of goodput is as used in Chapter 3 and expresses the bytes transmitted and ACKed at the sender, ignoring retransmissions. The routing protocol used is DSR.

For the simulation runs, the *s_increase_thresh* parameter is set to 4 and as such *cwnd* increases only every 5 ACKs, effectively limiting the increase rate to 1/5 of the original TCP Reno (and NewReno) algorithm. The *s_increase_thresh* variable was set to this value after several experiments with different values and having noted little effect in increasing the parameter further. This adjustment is titled the "slow" slow start modification of TCP (SS TCP) and its application to Reno and NewReno is named SS Reno and SS NewReno respectively. Figure 4.2(a) depicts the goodput of

---

[2]This is the size of the largest segment that the sender can transmit

Reno and SS Reno in string topologies as the hop count increases. The results in this case are mixed and the improvement in goodput not noteworthy in all cases. The most noticeable difference appears at 7 hops (8 nodes in the string topology) where there is a 6% increase in goodput whilst the worst case presents itself at 11 hops where there is a 4% decrease. For NewReno the equivalent results are included in Figure 4.2(b). In that case, the slow start modifications again provide mixed results with the best case noted at 8 hops (9 nodes in the topology), by exhibiting a 9% increase over the normal slow start procedure and the worst case indicated at 6 hops with a 4% decrease in goodput.

Algorithm 1 Slow start $cwnd$ increase

Require: $s\_increase\_thresh$ is the number of ACKs between increases, $s\_increase$ is initialised to 0

1: **if** $s\_increase = 0$ **then**
2:     $cwnd \leftarrow cwnd + 1$
3:     $s\_increase \leftarrow s\_increase + 1$
4: **else**
5:     **if** $s\_increase = s\_increase\_thresh$ **then**
6:         $s\_increase \leftarrow 0$
7:     **else**
8:         $s\_increase \leftarrow s\_increase + 1$
9:     **end if**
10: **end if**

The observed increase in goodput is explained as follows; as has been noted in [43] the segment drops caused by link contention due to hidden terminals become the main cause of segment loss in wireless networks when the hop count of the path is large enough. The SS Reno modification tries to reduce that effect by decreasing the increase rate of the slow start phase. For the sake of illustration, assume a given window, during the slow start phase, when a segment drop occurs due to link contention. Until that drop is noted by Reno (either through 3 dupACKs or an RTO), $cwnd$ keeps increasing exponentially due to ACKs returning to the sender for non-dropped segments. As extra segments are injected in the network and along the communications path, link contention is aggravated.

The routing protocol may, further, include a packet caching mechanism in the event of packet loss due to mobility. For instance, the DSR protocol [60] specifies the use of a maintenance buffer which contains dropped packets due to route changes. A packet that is discarded by the MAC mechanism may be considered to be dropped due to mobility and hence may end up in the maintenance buffer while the routing protocol attempts to "salvage" it by asking neighbouring nodes to provide an alternate route. Thus, it is possible that the packet is retransmitted at a later time, when the new route, which in this case is the same as the old one, is discovered by DSR. An example of this behaviour in both the case of DSR and AODV has been extensively discussed in Chapter 3. Such retransmissions from the routing protocol can cause even more link contention. Eventually, TCP may have to retransmit a lost segment either by restarting with the slow start phase in case of an RTO, or by entering the fast retransmit/fast recovery state in the case of duplicate ACKs. The SS modification forces TCP to not increase the sending rate as quickly during the slow start phase, and thus allows the segments to advance without adding greatly to spatial contention. A segment will eventually be dropped again after the optimal sending window has been reached but this event will happen at a later time than without the modifications.

Overall, though measurable, the performance advantages of the proposed modification are not great. The reason for this is that the slow start algorithm is not activated often enough to be effective. The slow start threshold value, which dictates when the slow start phase gives way to the congestion avoidance phase, is usually low in the examined scenarios and there is scarcely any time for the new method to make a difference. The time spent in the congestion avoidance phase is much longer than that spent in slow start and so the effects of the modification are not greatly noticeable.

4.3.2   Congestion avoidance modification (SCA TCP)

The congestion avoidance phase of TCP dictates a *linear* increase in the *cwnd* size per round-trip time (RTT). To achieve this, every ACK received acknowledging new data increases *cwnd* by $1/cwnd$ segments[3] [56].

In order to "slow down" this sending rate increase during the congestion avoidance phase a delay similar to the slow start modification presented above is introduced. The

---

[3]In actual TCP implementations, windows and segment sizes are measured in bytes and so the increment is $maxseg * \frac{maxseg}{cwnd}$, where $maxseg$ is the maximum segment size and $cwnd$ is expressed in bytes

proposed algorithm is shown in Algorithm 2. The behaviour of this modified congestion phase is dictated by the value of the *ca_increase_thresh* variable, which specifies
the level of delay added to the sending rate increase. Specifically, *cwnd* increases by a
full segment's worth every *ca_increase_thresh*+1 RTTs. The modifications are referred
to as "slow" congestion avoidance of TCP (SCA TCP) and their application to Reno
and NewReno are referred to as SCA Reno and SCA NewReno, respectively.

To evaluate the scope of improvement offered by these changes, we have conducted
experiments on string topologies by replicating the simulation setup in the previous
section. The *ca_increase_thresh* parameter was set to 4, which in turn implies that
the *cwnd* would grow only every 5 RTTs.

The results of this modification for a Reno TCP agent are shown in Figure 4.2(a).
The improvement in average goodput compared to plain TCP Reno is consistent and
ranges from 33-71% in the case of the string topology. The explanation for the increase
in goodput is the same as presented for the slow start modifications. However, the
slowdown of the increase rate in the congestion avoidance phase occurs far more often
than in the case of the slow start modifications in SS Reno, which introduces a more
substantial cumulative effect.

For NewReno the congestion avoidance modification results in a goodput increase
in the range of 32-71% for all path lengths. These are shown in Figure 4.2(b).

Algorithm 2 Congestion avoidance *cwnd* increase

Require: *ca_increase_thresh* is the no. of ACKs between increases, *ca_increase* is
   initialised to 0
1: **if** $ca\_increase = 0$ **then**
2:    $cwnd \leftarrow cwnd + \frac{1}{cwnd}$
3:    $ca\_increase \leftarrow ca\_increase + 1$
4: **else**
5:    **if** $ca\_increase = ca\_increase\_thresh$ **then**
6:       $ca\_increase \leftarrow 0$
7:    **else**
8:       $ca\_increase \leftarrow ca\_increase + 1$
9:    **end if**
10: **end if**

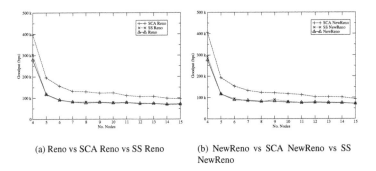

(a) Reno vs SCA Reno vs SS Reno   (b) NewReno vs SCA NewReno vs SS
                                              NewReno

Figure 4.2. Goodput vs length in string topologies for Reno and NewReno using SS
and SCA modifications

The following example illustrates the behaviour of the new mechanism and was encountered several times during simulation. Assume that the TCP sender receives 3 duplicate ACKs. TCP then fast retransmits the missing segment, halves *cwnd* and enters
the fast recovery phase. If the retransmitted segment reaches its destination without
triggering a congestion indication event, *cwnd* will increase at a rate of 1 segment per
RTT. By decreasing the *cwnd* increase rate to 1 segment per $ca\_increase\_thresh + 1$
RTTs, the TCP sender does not add to the link contention problem significantly and
is allowed to successfully transmit for longer around a large window close to the optimal one [43] than if *cwnd* were to be increased every RTT. The trade-off is that it is
possible to transmit around a sub-optimal window for a longer period as well, as the increase rate does not reach optimal size as quickly; however, this is offset by the longer
successful transmission of data around a larger optimal window. Another side-effect
is that it is possible to transmit with *cwnd* of size 4 or more for longer and as such
make use of dupACKs which can activate the fast retransmit/fast recovery algorithm
and recover quickly from segment losses.

It should be noted that the value of 4 for the $ca\_increase\_thresh$ parameter, as
introduced here, has been chosen only for illustration purposes. Section 4.4.1 includes
an exploration of an effective $ca\_increase\_thresh$ value.

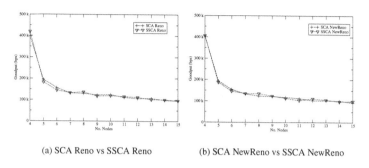

(a) SCA Reno vs SSCA Reno   (b) SCA NewReno vs SSCA NewReno

Figure 4.3. Goodput vs length in string topologies for Reno and NewReno using SSCA modifications

4.3.3   Cumulative modifications

As the SS and SCA techniques introduce changes at different phases of a TCP sender, it is useful to examine if the combination of the two methods may yield a cumulative improvement. To this end, we have combined the two techniques and produced an all-encompassing *increase_thresh* parameter with $s\_increase\_thresh = ca\_increase\_thresh = increase\_thresh$ for a TCP agent utilising both the SS and SCA methods.

To evaluate the effect of such a combination on the achieved goodput of a TCP sender, we have conducted experiments on string topologies as in the previous sections. Figure 4.2(a) shows the resulting average achieve goodput when the SS and SCA techniques are utilised in unison. This modification is titled SSCA and both the slow start and congestion avoidance increases are slowed down to 1/5 of the original Reno algorithm (i.e. *increase_thresh* is set to 4). The end result is not a cumulative increase in goodput but rather equivalent performance to the SCA slowdown technique. This is explained by the fact that the congestion avoidance phase dominates over the slow start phase and as such modifications in the latter do not significantly affect goodput. It is deemed, hence, superfluous to combine the two techniques and the SCA method remains in focus for the rest of this study.

Similarly to the Reno case, the mixed SSCA technique does not yield significant improvement over SCA for NewReno as can be seen in Figure 4.3(b).

Finally, note that the cumulative SSCA modifications as outlined in this section assume equality of $s\_increase\_thresh$ and $ca\_increase\_thresh$. Several experiments have been conducted in view of considering the case where the two thresholds are set at different values, thereby dictating different degrees of sending rate increase during the Slowstart and Congestion Avoidance phases. The resulting goodput in those cases is not significantly better than using the SCA method as described above; thus, the SCA method remains the focal point of this study.

4.3.4   Trace analysis of SCA TCP

This section shows the effect of the proposed SCA method changes in a string topology of 5 nodes (A,B,C,D,E). This discussion aids understanding and illustrates the workings of the SCA technique. Although this example uses TCP Reno as an agent the resulting conclusions are applicable to other Reno-based variants (such as TCP NewReno and SACK). The routing agent used is DSR. The other simulation parameters are as used in the previous sections.

For the purposes of this example and in the string topology, an FTP transfer is conducted from node A to node E (the end-nodes in the topology) and lasts for 100 seconds. The experiment is then repeated on the same topology with an SCA Reno agent with a threshold parameter of 4. As such $cwnd$ during the congestion avoidance phase is increased by a full segment every 5 RTTs and not every RTT as in the case of TCP Reno.

In Figure 4.4(a) the number of ACKed segments during simulation time is shown. At the end of the trial run SCA Reno has achieved significantly better goodput performance than Reno (approximately 38% improvement). It is also evident that the two variants perform comparably only for the first few seconds; after that time period (approximately 14 seconds) SCA Reno clearly outperforms Reno. Furthermore, the congestion window evolution during that time, shown in Figure 4.4(b), reveals that SCA Reno has maintained, a relatively small $cwnd$ throughout the connection time with a maximum of 5 segments in-flight at about the 65 seconds mark. In contrast, TCP Reno has maintained a larger $cwnd$, even reaching the value of 9 segments after 88 seconds.

A closer study of the first 20 seconds of the simulation time provides insight on the

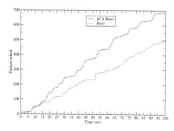

(a) Segments acknowledged over time -
after a few secs SCA Reno outperforms
plain Reno

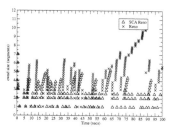

(b) Congestion window (*cwnd*) evolution - SCA Reno reaches smaller max.
cwnd than Reno

Figure 4.4. Behaviour of SCA Reno vs Reno

performance discrepancy. As can be seen in Figure 4.5(a) the two mechanisms (plain
and SCA Reno) behave identically during the initial slow start phase which lasts from
the beginning of the connection until *cwnd* (*cwnd*) takes the value of 7 at 0.6 seconds.
In this time slot the slow start mechanism, which is identical in both cases, is in effect.

Figures 4.5(b) and 4.5(c) depict the *cwnd* evolution during simulation time for
the Reno and SCA Reno method respectively. In those figures dropped segments are
denoted with a cross on the top bar of the graph. A segment drop event in this case
always refers to segments dropped by the routing agent due to the hidden terminal
effect. Notably, during the simulation no segment drops were recorded to have been
caused due to full buffer queues which verifies the findings in [43]. Also note that a
segment drop is not always accompanied by a drop in the value of *cwnd*, because the
routing algorithm salvages the segment and attempts to retransmit it later. From the
simulation trace it is apparent that the hidden terminal effect causes node A to drop
segment 5 at 0.21 seconds and declares a spurious route breakage. Soon after, out-of-
order segments appear at the receiving end of the connection. In this case, segment
6 is spuriously retransmitted at 0.8 seconds, and *cwnd* halves from 7 to 3 segments
as the fast retransmit/fast recovery algorithm is activated. Segment 6 was received
by node E at 0.776 seconds but due to segment reordering 3 dupACKs were sent to
node A (these were for segments 9, 7 and 10). Segment reordering can result from
the way the DSR maintenance buffer operates. According to the DSR draft [60] the

maintenance buffer holds segments for which a new route is being sought. In the case of the string topology as presented here there is only one route for every segment (route $A \rightarrow B \rightarrow C \rightarrow D \rightarrow E$). When the MAC layer gives up on transmitting a segment, the node will try to salvage the segment and discover the route again by enquiring neighbouring nodes for alternate routes to the destination. However, more segments might arrive and be put in the maintenance buffer in the mean time. If the queueing paradigm is Last-In-First-Out (LIFO), as in this case, segments may be forwarded out of order. The issue of segment reordering is relevant in the case of AODV [92] routing as well, where the local repair function can produce similar results.

After the 3 dupACKs both algorithms halve the *cwnd*. However, the classic Reno algorithm increases the *cwnd* 5 times more quickly during congestion avoidance than SCA Reno. In this case (at around the 1 second mark) it makes little difference as both algorithms experience a timeout due to segment 12 and both involve a *cwnd* of 3 (rounded down). A similar situation occurs at around 5 seconds. At that time spatial contention is high between nodes A and E but the slower congestion avoidance phase of the SCA algorithm has not had any impact yet. In order for a new segment to enter the network, *cwnd* must be increased by a whole unit and as such although the value of *cwnd* at around 5 seconds is 2.5 for SCA Reno and 2.9 for Reno it makes no difference. The advantage of the SCA algorithm is shown very clearly at 6-15 seconds. Plain Reno increases *cwnd* linearly and injects more segments into the network reaching up to 6 segments in flight. In contrast SCA Reno does not exceed 4 segments. Reno's extra segments are cached by the routing algorithm every time a route breakage occurs and are re-injected at a later time thus adding to the spatial congestion i.e. aggravate the hidden terminal effect. In the SCA case, there are fewer segments in the network and such effects are less pronounced after the inevitable timeout has been reached.

The basic idea of the proposed changes is that by restricting the growth of *cwnd*, there are fewer segments in flight for some period of time. As such, by the time the hidden terminal problem appears there will not be as many segments in the pipe to compete for medium access. As a consequence, the serious spatial link congestion as described in [43] is not as extreme as in the case of plain Reno. Subsequent transmissions and retransmissions do not have as many segments in contention for access the medium. Overall, both TCP versions exceed their optimal window at some point, only it is likely that this point is of a smaller value or at least the time spent at a non

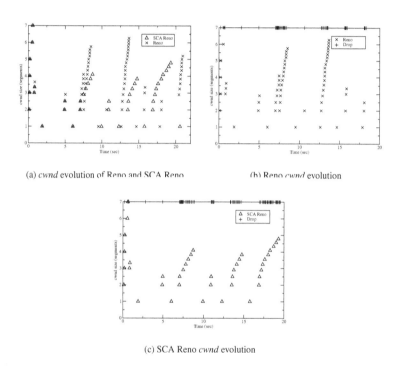

(a) *cwnd* evolution of Reno and SCA Reno

(b) Reno *cwnd* evolution

(c) SCA Reno *cwnd* evolution

Figure 4.5. *cwnd* evolution for the first 20 secs of simulation

problematic window zone is larger for SCA Reno. The previous statement implies that the time spent at smaller than optimal values is longer, as well, but this is more than offset by the fact that when a failure occurs at a high *cwnd* value, as is the case with TCP Reno, the spatial congestion is serious and is augmented by retransmissions from the maintenance buffer of the routing protocol.

4.4 Evaluation of SCA TCP

In this section, the SCA technique is evaluated thoroughly and is further contrasted to a different approach presented in the literature which deals with the effects of spatial contention on TCP, namely the adaptive Congestion Window Limit (CWL) method [26].

4.4.1 Performance analysis of SCA TCP

Before a performance comparison of SCA with the adaptive CWL and other TCP strategies can be attempted, it is necessary to evaluate the effect of different SCA parameter ($ca\_increase\_thresh$) values on the performance of the SCA method when applied to TCP. Such a process would facilitate the setting of a "generally good" default parameter which will be used in the subsequent performance evaluation. The next section outlines the results of this procedure and an evaluation in general mobility scenarios follows.

Identifying a default SCA parameter

To identify an adequate $ca\_increase\_thresh$ parameter for the SCA method we have performed simulation experiments for different string topologies and noted the resulting goodput.

Specifically, string topologies of various length have been considered, ranging from 3 to 15 hops (4 to 16 nodes). String topologies of less than 4 nodes have not been evaluated as hidden terminals due to interference are not evident in such topologies [107]. For each simulation run, an FTP connection between the end-points is initiated at the beginning and lasts until the end of the simulation at 900 secs. Per string topology, a

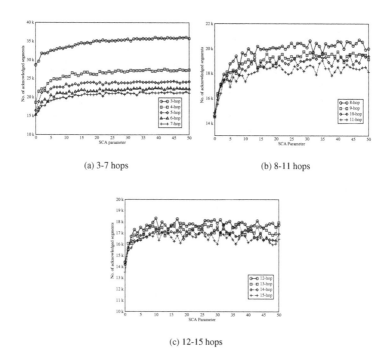

(a) 3-7 hops   (b) 8-11 hops

(c) 12-15 hops

Figure 4.6. Number of transmitted segments vs SCA parameter in a string topology of 1 to 15 hops using DSR

different SCA parameter was applied, ranging from 0 to 50, and the goodput at the end was recorded. Note that an SCA parameter of 0 denotes plain Reno, i.e. no modifications to the Reno congestion avoidance algorithm.

Figure 4.6 shows the result of the simulation runs for DSR. Notably, the number of transmitted segments for each hop count flattens or stabilises after the SCA parameter has reached the value of 10. This in turn implies that further possible improvement shown in the simulation metric is not significant beyond that value. This observation is true for all the hop counts considered here (3-15 hops) and, hence, regardless of the string length, the choice of 10 as the value for the SCA parameter would appear to lead

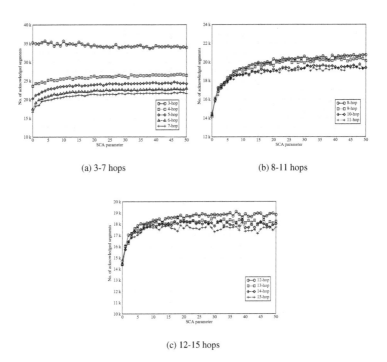

(a) 3-7 hops   (b) 8-11 hops

(c) 12-15 hops

Figure 4.7. Number of transmitted segments vs SCA parameter in a string topology of 1 to 15 hops using AODV

to overall "good" performance.

The same behavior is evident in the graph in Figure 4.7 for the AODV routing protocol. Specifically, at around the 10 mark point, any further increase in the SCA parameter does not yield significant improvement in the performance metric.

For clarity, it should be noted that when considering TCP performance in both Figures 4.6 (DSR) and 4.7 (AODV), emphasis should be placed on the *trend* evident rather than the individual goodput values. Hence, in both cases it is of interest to note an approximate SCA threshold where the graph "flattens" rather than note global "peaks" where goodput is at a maximum. Further in this chapter (in Section 4.5.2) an enquiry is made into optimising the SCA technique by setting its parameter according path length; in that case, the parameter is set according to these global peaks rather than the trend of the goodput graph for each hop count.

Finally, note that it is quite possible to make use of feedback from the DSR protocol or any other route length aware routing protocol so that the SCA mechanism is deactivated if the hop count is less than 4. This would offset the issue of activating the SCA modifications when it is not needed.

Notes on the choice of SCA parameter

The choice of a default SCA parameter, as performed in this section, may not lead to an optimal choice for every topology and mobility pattern, since different spatial reuse characteristics may be evident in a given path depending on the distance between successive nodes [26]. Nonetheless, the choice presented herein represents a "good" value for all path lengths as exhibited in typical string topologies and is expected to perform well, if not optimally, for other path types. This assumption is widely shared by previous related research [25, 31, 105] and is explicitly stated here. As such, adopting a formal approach on evaluating the precise effect of altering the SCA parameter, through parametric sensitivity analysis, would reveal a scenario specific optimisation but would provide little insight for the general case. Overall, it should be noted that the performance measurements in dynamic topologies presented in the following sections indicate that the choice of 10 as the default SCA parameter does result in significant goodput gains.

Finally, it should also be stated that the choice of 10 as an SCA parameter represents a deliberately conservative decrease in the congestion window increase rate. Even though a larger SCA value may result in higher gains, such a value may negatively and severely impact TCP convergence, responsiveness and its ability to utilise its fair share of the bandwidth when interacting with other flows [57]. A brief enquiry into those issues is included in Section 4.4.2.

Simulation setup

To validate our choice of a default SCA parameter (set to 10), we have evaluated the performance of SCA Reno against plain Reno and Vegas over DSR and AODV in

dynamic MANETs. The simulation results indicate whether the SCA strategy aids
Reno in achieving Vegas-like performance or better, i.e. if it functions as intended, in
dynamic scenarios.

Simulation area and mobility model: The simulation area is defined to be a strip
of 1500x300m where 50 nodes are placed randomly. To create results comparable to
previous TCP literature studies [19,25,34] node pause times of 0 and 40 seconds have
been considered and maximum node speeds of 2, 5 and 10 m/s. Mobility has been
simulated using the random waypoint mobility model over 50 different mobility scenarios for each pause time/mobility combination. Each trial run lasted for 900 seconds.
To ensure fairness in the results, the same topologies were used for the different TCP
agents over the same pause times.

TCP transfer setup and metric used: In each simulation run, a TCP connection
is set up between two randomly selected nodes and an FTP transfer session was initiated for the duration of the simulation. A maximum window size of 64 is chosen
for both the congestion (*cwnd*) and advertised (*awnd*) windows. During the course of
the experiments the maximum sending window size (=min$\{cwnd,awnd\}$) was never
reached and as such our performance metric was not limited by that bound. The performance metric measured in the simulation experiments is goodput and the average
of the 50 topology results for each pause time/mobility combination is considered.
Welch's t-test has been performed on the observations to determine if there are statistically significant differences in the performance of the TCP agents and 90% confidence
intervals have been computed; however, these are not included in the graphs to avoid
cluttering.

Notably the retransmission timer's maximum value was set to 240 seconds as recommended in RFC 1122 [15]. Other work in the literature which examines a different
technique to deal with the issue of TCP and spatial contention [25] has used a lower
maximum RTO value. However, the authors in [25] have acknowledged that their
imposed RTO limit of 2 seconds is low, but chosen so as to minimise the time the
connection spends idle when a broken communications path may already have been
re-established by the routing mechanism. However, we believe that a realistic outlook
of typical TCP performance in MANETs has to maintain parameters in TCP as the

standards recommend, which ensures that the congestion control mechanism functions as intended.

In these simulation runs five TCP variants have been evaluated; plain Reno, Vegas, and three SCA agents with *ca_increase_thresh* (or SCA) parameters set to 1, 5 and 10. The value of 1 is significant as it denotes the initial impact of the SCA mechanism (i.e. the impact of activating it with the lowest possible parameter). The value of 10 denotes how a "default" set agent (as determined by simulation in the previous section) would behave. Finally, the value of 5 was chosen as it represents the mid-point between the optimal and minimum parameters and its results allow for interesting observations in the subsequent discussion.

Finally, we have considered AODV and DSR as routing protocols with parameters as used in Chapter 3 and as detailed in Appendix A.1. The simulation parameters with respect to the signal propagation model and wireless transceiver setup are the same as the ones used in Chapter 3 and throughout this chapter.

Results and discussion

The results under DSR routing for 0 seconds (continuous mobility) and 40 seconds pause time are presented in Figure 4.8. For maximum node speeds of 2, 5 and 10 m/s, SCA Reno with parameter 10 outperforms plain Reno by 4%, 26% and 24% respectively for the topologies with 0 seconds pause time. Similarly, for 40 seconds pause time SCA Reno with parameter 10 improves over plain Reno by 5%, 6% and 15% for maximum node speeds of 2, 5 and 10m/s respectively. The decrease in improvement compared to continuous mobility scenarios is largely attributed to the random source/destination pairs which are chosen for each scenario. In certain scenarios, partitioning ensued in the network and it became impossible for the routing algorithm to find an alternate route and as such the SCA method could not "improve" upon the plain Reno variant.

Expanding on the last point, an interesting situation was observed during the experiments. The SCA technique did lead to RTOs due to spatial contention similar to the case of plain Reno, with the difference that SCA maintained fewer segments in the pipe on average. Notably, the RTOs occurred, generally, at a different time interval than plain TCP Reno. If an RTO occurred at an "inconvenient" time, such as when network partitioning was about to occur, TCP performance suffered under SCA compared to Reno, assuming Reno RTOs occurred at more "convenient" times. We believe that the adverse effect of RTOs on TCP performance due to interference/hidden terminals should not be understated and therefore neither took special precautions to avoid "badly timed" RTOs nor ensured that at the end of an RTO the routing path would be known to the source and ready for use.

In general, although setting the SCA parameter to 10 consistently outperforms the SCA Reno with SCA parameter of 1 for all pause times, it is closely matched at certain pause time/node speeds by SCA Reno with parameter of 5 and can even present slightly worse results as in the cases of 10m/s speed at 0 seconds pause time and 5m/s speed at 40 seconds pause time. Such behaviour is somewhat expected as the performance of the SCA parameter of 5 in Figure 4.6 suggests that it is a competent contender for different hop counts but on average should be worse than the SCA parameter 10 agent under different path lengths, which is what can be observed in this case.

When noting the performance of SCA against Vegas, it is evident in the case of DSR that the difference in performance is only significant in a few cases, specifically at 10m/s for 40 seconds pause time where it is about 7% and at 5m/s for 0 seconds pause time where it is about 4%. At all other pause time/node speed combinations the discrepancy in performance is either very small (about 1-2%) or not statistically significant. Nonetheless, this is a clear indication that SCA achieves similar performance levels to Vegas, without incurring the computational overhead of the latter [16].

Figure 4.9 depicts the performance measurements of SCA Reno under the AODV routing protocol. The trend is similar to that of DSR, namely improved goodput ranging from 4%-10% for 0s pause time and 6%-8% for 40s pause time over plain Reno. As is the case with DSR routing, using 10 as the SCA parameter results in the highest and most consistent improvement over plain Reno in our simulations when compared to the parameter settings of 1 and 5. The decrease in improvement compared to DSR routing is attributed to the faster recovery of AODV from "false" route breakages (which reduces the impact of consecutive RTOs). Instead of asking neighbours for alternate routes, AODV actively searches for a new route by initiating a local route discovery process from the point of failure. The new route can be quickly re-established even in the case of a real route breakage as the route to the destination is likely to be easily rediscovered through other neighbouring nodes close to the point of failure.

The simulation results in the case of Vegas indicate a similar pattern to that of DSR; Vegas is only outperformed in a few cases (4% at 10m/sec for 0s pause time and 4% at 5m/s for 40s pause time) whilst the two methods are equivalent otherwise. Overall, the simulation results indicate that SCA Reno outperforms Reno in a single connection environment under both the DSR and AODV routing protocols while achieving equivalent performance to Vegas.

To understand the performance of SCA Reno over a somewhat denser network traffic environment we have also conducted experiments using multiple TCP flows with the same 50 topologies. Specifically, five TCP connections were initiated for 900 seconds, carrying FTP traffic. The communicating pairs were randomly chosen for each scenario and the performance metric measured was the same as before. The results of the aggregate goodput are presented in Figure 4.10 for 0 and 40 seconds pause time and different node speeds for the DSR routing protocol. The improvement in goodput over plain TCP Reno ranges from 3-12% for 0 seconds pause time and 4-12% for 40 seconds pause time. Note that our SCA solution (not unlike the adaptive CWL solution examined in the next section) is not meant to solve the interference problem among different flows but improve performance when the interference is caused by the flow "onto" itself. Under the AODV routing protocol the performance increase is not as noticeable as in the case of DSR for multiple connections as can be seen in Figure 4.11. Specifically, the performance improvement over TCP Reno ranges from 2-5% for 0 seconds pause time and 2-4% for 40 seconds pause time. This is attributed to the ability of AODV to quickly recover from false route breakages and the fact that other factors affect TCP performance such as network partitioning which leads to repeated RTOs. In the case of network partitioning the faster congestion avoidance phase of Reno can be beneficial as it can inject more segments on the network for the time the route is valid compared to SCA Reno. Nonetheless, SCA Reno deals with the effects of spatial contention much better than plain TCP, and that results in a slight advantage for SCA Reno.

Note that in this case the performance of SCA compared to Vegas is again comparable, with notable exceptions being the points at 10m/s for 0s and 40s pause time in the case of DSR (where the difference in goodput is 7% and 5% respectively in favour of SCA) and 10m/s for 0s pause time in the case of AODV (where the difference is approximately 5%). The performance increase achieved by SCA in scenarios

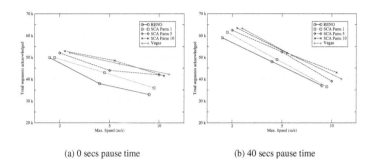

(a) 0 secs pause time   (b) 40 secs pause time

Figure 4.8. Goodput of a single TCP flow vs maximum speed in dynamic topologies using DSR

with multiple flows provides some indication that the SCA modification functions as intended even between flows although we do not claim this is to be true in every scenario as the mechanism deals with intra and not inter-flow interference.

4.4.2 Performance comparison of SCA TCP and adaptive CWL

In this section, we evaluate the effectiveness of the SCA TCP strategy against an existing technique in the literature aiming to alleviate the throughput-reducing effects of spatial contention, namely the adaptive CWL setting strategy. Since both the SCA and adaptive CWL strategies are applicable to path length aware routing protocols, both are evaluated under DSR routing. Invariably, the adaptive CWL strategy cannot be used with a path length agnostic routing agent in contrast to the SCA technique, which requires no such coupling of the routing protocol with the transport agent. The next section introduces the adaptive CWL strategy and is followed by a performance evaluation comparison and discussion.

Congestion Window Limit (CWL) method

The Congestion Window Limit (CWL) approach enforces a restriction on the maximum congestion window (*cwnd*) of TCP, so as to maintain few outstanding segments in the pipe at any one time and minimise spatial contention. This spatial contention is

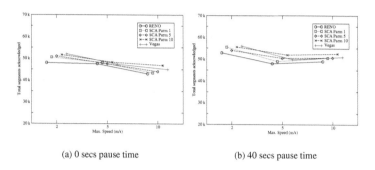

(a) 0 secs pause time

(b) 40 secs pause time

Figure 4.9. Goodput of a single TCP flow vs maximum speed in dynamic topologies using AODV

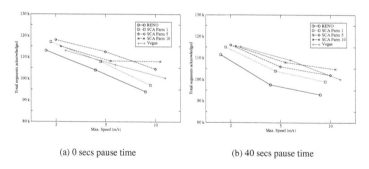

(a) 0 secs pause time

(b) 40 secs pause time

Figure 4.10. Aggregate goodput of 5 TCP flows vs maximum speed in dynamic topologies using DSR

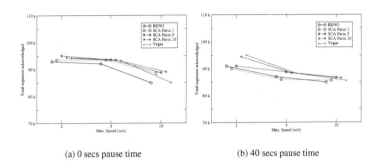

(a) 0 secs pause time   (b) 40 secs pause time

Figure 4.11. Aggregate goodput of 5 TCP flows vs maximum speed in dynamic topologies using AODV

caused by the flow onto itself and its main cause is, thus intra-flow interference. As such, the method addresses the same issue as the SCA technique and thus is similar in research scope.

An outline of the method follows. TCP maintains for each connection a *cwnd* state variable which throttles the sending rate at the sender. Moreover, the sender receives flow control information from the receiver, which is maintained in a sender's advertised window (*awnd*) state variable for each connection. The amount of data that can be outstanding in the network at any one time is set by the $sending\_window = max\{cwnd, awnd\}$ [98]. A limitation imposed on the value of *cwnd* implies a limitation on the sending window and as such the number of outstanding segments that can be present in the pipe by the TCP sender. The authors in [107] and [43] have demonstrated that imposing a limit on the maximum sending/congestion window leads to improved throughput as fewer segments contend for access to the medium at any one time leading to better spatial reuse of the medium. As such, fewer link layer drops occur due to interference which leads to less spurious route breakages and TCP retransmission timeouts (RTOs). A technique following the above paradigm is labelled as a CWL method [25].

Work by Chen et al. has demonstrated a mechanism that dynamically adjusts the CWL according to the path length [26]. Instead of relying on analytical estimates of

Table 4.2. Limiting the maximum congestion window (*cwnd*)

Hop Count(h)	Max. *cwnd*
$h \leq 2$	2
$2 < h \leq 4$	1
$4 < h \leq 6$	2
$6 < h \leq 10$	3
$10 < h \leq 13$	4
$13 < h \leq 15$	5
$h > 15$	not obtained

optimal values for *cwnd* (as done in [43]), the authors have instead opted for discovery through simulation of the optimal CWL value for several path lengths as shown in Table 4.2. This *adaptive* CWL strategy receives path hop count information provided by the DSR protocol to set the maximum *cwnd* value of the TCP agent. These changes have been shown to improve throughput 8-16% over TCP Reno in multiple flows scenarios. In this work we evaluate the performance of the adaptive CWL method as described in [25], which supersedes the original fixed CWL method [108].

Evaluation results of SCA TCP vs adaptive CWL

To evaluate the effectiveness of the SCA technique against the adaptive CWL strategy we have conducted further simulations. The simulation parameters and experimental setup are identical to the ones used in the previous SCA TCP evaluation (Section 4.4.1). In all experiments, the SCA parameter for SCA Reno was set to the default value of 10. The hop count values considered for dynamically adjusting the congestion window using adaptive CWL are the same as in [25] and are presented in Table 4.2. Measurements were also taken of a plain Reno implementation to be used as the baseline for comparison. The routing protocol used in the evaluation was DSR so as to make our results comparable to the ones in the adaptive dynamic CWL work [25].

The goodput results for a single TCP flow appear in Figure 4.12. SCA TCP improves upon the performance of the adaptive CWL strategy by 4-9% and 7-20% for 0 and 40 seconds pause time respectively for all maximum speed scenario settings. The improvement offered over Reno ranges between 4-26% for 0 seconds pause time and 5-15% for 20 seconds pause time across different node speeds. The adaptive CWL strategy maintains a goodput average that is better than Reno but is in some negligible (2m/s at 0 seconds pause time and 5m/s at 40 seconds pause time) and it even proves worse than plain Reno by 4% at 2m/s for 0 seconds pause time.

In our experiments with single flows the CWL method performed comparably to SCA at some topologies and much less than optimally at others. The apparent discrepancy of our results to the ones presented in [25] is mostly attributed to the more accurate setting of the maximum RTO in our experiments. The maximum RTO setting was set to 240 seconds as recommended for Internet hosts [15] instead of 2 seconds as in [25]. Setting the maximum RTO to 2 seconds does improve the reaction of TCP to route re-establishment as the maximum time a (re-established) route may remain unnecessarily non-utilised is close to 2 seconds. However, as noted before, this setting is not recommended for interaction with Internet hosts [15]. In fact, by their own admission in [25] the authors identify their choice of value for the maximum RTO as "small" and not recommended for widespread use. However, such a modification understates the effect of RTOs and skews the end results.

To understand the effect of limiting the maximum RTO to a small value consider the following scenario. The SCA technique maintains on average more segments in flight than adaptive CWL as it is not limited by a *cwnd* bound. Although some segments are lost due to spatial contention in SCA, several of these are salvaged by DSR which retransmits them at a later time and can trigger a 3 duplicate ACK (dupACK) response from the receiver which activates the fast retransmit/fast recovery phase of TCP at the sender. DupACKs are produced either because the segments arrive at the receiver out-of-order, or because there are "holes" (i.e. non consecutively sequenced segments) in the destination's receiving buffer. The adaptive CWL method does not need the dupACKs heuristic as it is designed to avoid segment drops due to spatial contention. However, if the route breakage is due to mobility the dupACKs heuristic can be a valuable tool and enable TCP to recover quickly from the route breakage. When the fast retransmit/fast recovery mechanism is in effect TCP immediately retransmits the segment "known" to be lost without waiting for an RTO. This quick retransmission and utilisation of a possible re-established route is impossible in the case of the

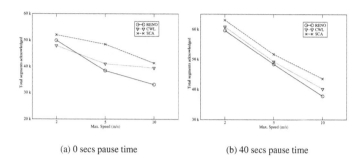

(a) 0 secs pause time   (b) 40 secs pause time

Figure 4.12. Goodput for a single TCP flow vs maximum speed in dynamic topologies using DSR (SCA vs adaptive CWL)

adaptive CWL strategy when the maximum *cwnd* is smaller than 4 segments (which is true for paths shorter than 13 hops, as shown in Table 4.2). Hence the adaptive CWL technique has to rely on the expiration of the RTO timer for segment retransmissions. Furthermore the exponential RTO backoff aggravates the problem as it can lead to several seconds of inactivity, even if the route has been restored in the mean time (an analysis of this phenomenon is included in [49]). If the RTO is set to a small enough value, as in [25], this aspect of the problem is ignored and the performance evaluation results are deceptively favourable.

To obtain insight on the effectiveness of both strategies with multiple flows, although neither was particularly targeted to deal with the issue, we have applied both techniques to concurrent TCP flows in MANET environments. The number of FTP flows (carried with TCP) was set to 5 with random source/destination pairs. The simulation setup parameters are the same as in the multiflow SCA evaluation in Section 4.4.1 and are not repeated here for brevity.

The results of this evaluation appear in Figure 4.13. The SCA technique again proves superior to adaptive CWL (in terms of aggregate goodput) for all maximum speeds and at all pause times and at a margin of 2-10%. From these results it may be suggested that the SCA technique maintains its advantage over adaptive CWL under some multiflow network traffic.

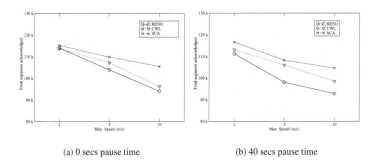

(a) 0 secs pause time   (b) 40 secs pause time

Figure 4.13. Goodput for 5 TCP flows vs maximum speed in dynamic topologies using DSR (SCA vs adaptive CWL)

Limitations of adaptive CWL

Although the adaptive CWL method may alleviate spatial contention in single flows it can be shown to have serious problems competing with plain TCP flows. To illustrate this, consider the following scenario as described below.

We assume a string topology with 5 nodes and two FTP flows that have the same source and destination, specifically the end-nodes of the string topology. The simulation lasts 900 seconds and the TCP flows carrying the FTP traffic are named flow A and B. Flow B uses the plain Reno algorithm whilst flow A uses the adaptive CWL adjustment. In this case the maximum $cwnd$ is set to one segment as the route length is fixed to 4 hops (Table 4.2). DSR routing is used as the routing protocol and all the simulation parameters are as in the previous section.

The grey shaded area in the graph in Figure 4.15 represents the total number of segments that have been transmitted by the sender and successfully acknowledged by the receiver (i.e. goodput) for the adaptive CWL flow as a function of time. In this case, it is particularly noticeable that the Reno flow utilises much more than its fair share of the bandwidth. Tracing reveals that during the simulation run the maximum $cwnd$ of one set by the adaptive CWL strategy leads to several RTOs for flow A because each segment loss cannot be recovered through the fast retransmit procedure as there are not enough segments in the pipe to utilise the dupACK heuristic. Flow B (Reno) is not restricted on the amount of segments it can inject in the pipe though, and has several segments at any one time competing with flow A's single segment for spatial usage of the medium. Hence, it becomes very likely that the single segment of flow A is dropped which leads to an RTO in the adaptive CWL sender before another segment is transmitted. The RTO timeout value doubles for each consecutive segment loss which further aggravates the situation for flow A. At the end of the simulation the average *cwnd* size for flow B is 7.3 whilst for flow A it is 1 (the maximum value).

Figure 4.14 shows the goodput results (in total segments) for the same topology when the adaptive CWL agent of flow A is replaced with other types of TCP agents. Essentially the same experiment as described above is executed with flow B always being a Reno agent and flow A being one of the following TCP variants: Reno, SCA Reno with parameter 5 (SCA 5), SCA Reno with parameter 10 (SCA 10), a dynamic CWL agent (CWL) or an agent with a fixed congestion window limit of $2\ldots6$ segments (cwnd $2\ldots6$). The simulations are run under both DSR and AODV to demonstrate the universality of this discussion. The Reno/Reno interaction ensures fair use of bandwidth for both flows. The SCA/Reno flow pair results reveal that the Reno agent "steals" some of the SCA bandwidth, which is expected as the SCA sender is not as aggressive during the congestion avoidance phase as Reno. In fact, in this case, fairness is an especially poignant issue in the adaptive CWL case where flow B (Reno) utilises $\frac{8}{9}$ of the available bandwidth whilst flow A (adaptive CWL) makes use of the rest (for the SCA Reno/Reno case, the ratios are $\frac{1}{3}$ and $\frac{2}{3}$ for the respective flows). Increasing the maximum *cwnd* size to 6 improves fairness as is apparent in Figure 4.14 because the behaviour of plain Reno, which has an average *cwnd* size of 5.68 in the Reno/Reno scenario, is emulated as the maximum *cwnd* size increases.

The time (in seconds) spent in each TCP phase for certain flows (Reno, SCA with parameter 10 and CWL) is shown in Table 4.3. Note, that the SCA agent experiences less overall RTO time than Reno and spends more time in the congestion avoidance phase, i.e. keeps transmitting for longer than Reno. Furthermore, for both Reno and SCA Reno, the congestion avoidance algorithm is active for most of the duration of the simulation run. This observation has been noted several times in this dissertation and it is the reason the SCA modification on the sending rate increase is applied to the congestion avoidance rather than the slow start phase.

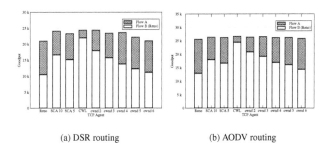

(a) DSR routing   (b) AODV routing

Figure 4.14. Shared bandwidth between two flows in a 4-hop string topology

Table 4.3. Time spent (in secs) in each phase for different TCP flows

Phase	Reno	SCA Reno	adaptive CWL
Slow start	54.8	9.7	1000
Congestion avoidance	842.9	957.1	0.0
RTO	102.3	33.2	0.0

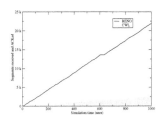

Figure 4.15. Flow share between plain Reno and adaptive CWL agents in a 5-node string topology

4.5   Other considerations

The following section contains discussion on the implications of the SCA method introduced, with respect to inter-flow spatial contention/interference (i.e. interference evident between separate flows) and feedback from the routing protocol. Each topic is discussed in turn.

4.5.1   Multiple flows

In this section we evaluate the impact of the SCA method and its different parameters as applied to multiple connections between two nodes. Overall, the study of inter-flow interference and spatial contention at the transport layer level can only be conducted on multiple flows that use the same route as these are the ones the transport agent is directly aware of. It is not possible for a TCP agent (assuming that the end-to-end paradigm remains intact) to realise the number of other flows in the network sharing the same path, unless feedback is drawn from some other source, like the routing protocol. It is, however, possible from the end-to-end perspective of TCP to note the effect of other flows on available bandwidth and make inferences about their interaction. Previous work on Internet transport dynamics has made it possible to infer the fair share of bandwidth that a connection should utilise by using RTT measurements or by appraising ACK feedback [17, 22]. Such mechanisms, however, do not apply directly to *ad hoc* networks because of fundamental differences in the access mechanism of the shared medium [26, 43].

The following analysis is based on the assumption that when several TCP connections are established in a {source, destination} pair, the corresponding flows share much of the same path. By conducting a series of experiments on string topologies of different length it is possible to isolate the effect on goodput of utilising different parameters for the SCA technique and make use of the resulting observations on dynamic topologies. By noting the optimal SCA parameter per hop count, for a given number of TCP connections that facilitate communication in the same {source, destination} pair, it is then possible to tune the agents so that goodput is optimised.

Prevalent single path routing protocols like AODV [92], DSR [60] and OLSR [28], which have been used throughout this dissertation, guarantee that multiple flows of a

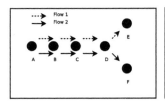 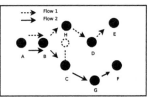

(a) Original flow path   (b) Flow path after topology changes

Figure 4.16. Common hops for different flows

{source, destination} pair will utilise the same path. As such, there is a large scope of applicability for the method proposed. Notably, throughout this work the semantics of the transport layer are kept intact, i.e. the proposed enhancements only require changes at the end-points and specifically only at the sender. It should be noted that the discussion that follows does not apply in the case of multipath routing [74] as the {source, destination} pair does not guarantee a common path.

Another special consideration is evident in the case of source routing and is explicitly discussed here. In the case of such protocols (like DSR [60]) , it is, in fact, possible for the TCP agent through interaction with the routing protocol to be aware of flows that share much of the same path even though they belong to different {source, destination} pairs. This holds true as long as the flows share the same source. Figure 4.16 depicts such an example, where TCP flows 1 and 2 share much of a common path, namely $A \to B \to C \to D$ and even exhibit the same hop count. In this case, the two flows interfere with each other and since node A is aware of the existence of both (being the originator of both), it can tune their SCA parameters so as to optimise their combined goodput. However, if some kind of local route repair mechanism is used, it is possible for the routes to diverge without any notification to the source. In such a case, node A will be making decisions based on outdated information which might have a negative impact on goodput. In the case of flows with the same source and destination targets, no such case is possible; these cases are the focal point of the following discussion.

Determining a suitable SCA parameter

In a series of experiments we attempt to approach empirically and through simulation the optimal parameterisation of the SCA NewReno method according to the path hop count. The simulation parameters are identical to the ones used in the SCA evaluation (Section 4.4.1).

Simulation scenarios are set as follows; in string topologies of various lengths (1-15 hops) FTP connections are established between the end-points (meaning the same {source,destination} pairs). These connections are facilitated by TCP agents with different SCA parameters, ranging from 0 to 50. The simulation lasts for 900 seconds and the aggregate goodput is recorded. Note that an SCA parameter of 0 denotes plain NewReno, i.e. no modifications to the NewReno congestion avoidance mechanism. This setup will help identify a suitable SCA parameter for multiple flows (the one exhibiting the highest goodput) which can then be set as the default value.

Figure 4.17 shows the goodput performance of 2 parallel SCA NewReno flows along string topologies of different length. Notice that there is no performance improvement in the case of 1 and 2-hop string topologies as the hidden terminal effect cannot occur in such short strings and spatial contention is minimal. For 3-hops there is some goodput improvement (up to 19%), but it is limited, since the SCA mechanism only helps alleviate hidden terminal effects that derive from the transmissions of ACKs from the destination (the 4th node in the string; the 1st node is the hidden terminal in this case). Diminishing the sending rate increase when there is no effect from interference (for path length smaller than 3 hops) can lead to a performance penalty, especially in the presence of background traffic. However, as in the case of a single flow examined earlier, it is feasible to make use of feedback from DSR or any other route length aware routing protocol so that the SCA mechanism is deactivated if the hop count is less than 3. Such information (when available) offsets the issue of activating the SCA modifications during the sending rate increase if it is not necessary, although the impact of false activation is not great in the case of low traffic.

There is a significant performance improvement for the hop counts examined, similar to that noted in single flows using SCA in Section 4.4.1. The number of transmitted segments for each hop count noticeably flattens or stabilises after the SCA parameter reaches the value of 25. This, in turn, implies that further possible improvement in the

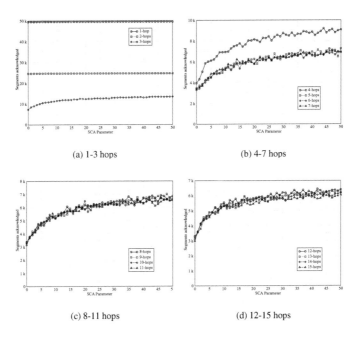

(a) 1-3 hops

(b) 4-7 hops

(c) 8-11 hops

(d) 12-15 hops

Figure 4.17. Aggregate goodput and SCA parameter in a string topology of 1 to 15 hops with two TCP flows

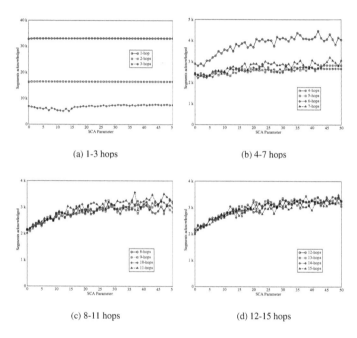

(a) 1-3 hops

(b) 4-7 hops

(c) 8-11 hops

(d) 12-15 hops

Figure 4.18. Aggregate goodput and SCA parameter in a string topology of 1 to 15 hops with three TCP flows

simulation metric is not significant beyond that value. The above observation applies
to all the hop counts considered here (3-15 hops). Hence regardless of the string length
the value of 25 would appear to be effective (in goodput terms) as an SCA parameter.

We have also empirically approached the discovery of a sufficient default SCA
parameter for up to 7 flows. For 3 flows the plateau of improvement for goodput is
found again to be for an SCA parameter of 25, as shown in Figure 4.18. For more
than 3 flows the impact of SCA on goodput is minimal mostly because there is heavy
spatial contention for any *cwnd* larger than 1. Essentially, there is insufficient time
for the *cwnd* evolution to allow the SCA method to be useful since any *cwnd* increase
beyond a single segment causes spatial contention.

Performance evaluation

To validate the results of the previous section and examine the choice of a default SCA
parameter per number of simultaneous TCP flows, we evaluate, the performance of
SCA NewReno against plain NewReno over DSR in dynamic MANET environments.

Simulation setup: The simulation parameters are set as follows: The simulation
area is 1500x300m where 50 nodes are placed randomly. The simulation model, signal
propagation characteristics and node configurations are as set as in the single flow SCA
evaluation (Section 4.4.1). The mobility model considered is the random waypoint
model with pause times of 0 and 40 seconds and mean node speeds of 1, 2, 5 and 10
m/s.

In each simulation run, multiple FTP connections are set between two randomly
chosen nodes at the beginning of the simulation and the TCP connections facilitating
the data transfer begin transmission. The FTP connections, and thus the TCP flows, are
active throughout the simulation time, which is 900 seconds. The performance metric
measured in the simulation experiments is goodput (in segments), averaged between
the TCP flows. In these simulation runs the SCA NewReno with SCA parameter of 25
(the default) has been evaluated against plain NewReno.

Discussion: The goodput results for 2 TCP flows and for 40 seconds pause time
are presented in Figure 4.19(b) for 2 parallel TCP flows. For mean node speeds 1,

2, 5 and 10 m/s SCA NewReno outperforms plain NewReno by 12%, 11%, 8% and 4% respectively. As the mean node speed increases, the performance improvement decreases as routes remain stable for less time on average due to increased mobility. The explanation for the decrease is as follows; the SCA Reno mechanism needs time to activate as its effects take place after the slow start mechanism. The purpose of the slow congestion avoidance phase of SCA TCP is to provide the pipe with additional time to resolve its spatial contention burden before injecting an extra segment. However, when the environment is highly dynamic, established routes are very ephemeral and the requirements of the SCA technique are not met, which in turn means that the improvement in performance is not as apparent.

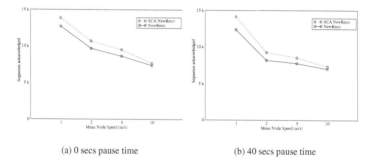

(a) 0 secs pause time   (b) 40 secs pause time

Figure 4.19. Average goodput for two TCP flows using NewReno and SCA NewReno

Similarly in continuous mobility environments, the improvement in goodput of SCA NewReno over plain NewReno is 4-9% for mean node speeds of 1, 2, 5 and 10m/s as shown in Figure 4.19(a) for 2 TCP flows. The improvement is significant under such conditions but not as prominent as in the case of 40 secs pause time. The reasoning for this drop in improvement lies in the fact that, similarly to a decrease in mean node speed, an increase in pause time favours the creation of longer lived routes than otherwise.

The goodput results for 3 simultaneous TCP flows in the same topologies are included in Figure 4.20. The improvement in goodput ranges from 7-12% and 5-8% for

40 and 0 seconds pause time respectively. The previous observations on the discrepancy in improvement over different pause times for two flows are valid for 3 flows as well. For more than 3 simultaneous flows our experiments have shown that the SCA method has no discernible effect on goodput. As mentioned in the previous section, the optimal *cwnd* for TCP under such conditions is around one segment and it is not possible for the SCA mechanism to function properly. This is not an issue exclusive to the SCA method; multiple segments in-flight (as a result of multiple TCP flows) along sufficiently long paths would cause deterioration in TCP performance regardless of the TCP agent used as there would be too many segments in the pipe at any one time for spatial reuse to be exploited [25, 26, 43, 72].

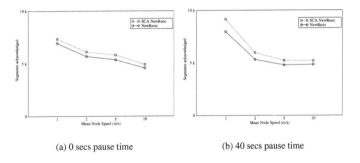

(a) 0 secs pause time   (b) 40 secs pause time

Figure 4.20. Average goodput for three TCP flows using NewReno and SCA NewReno

4.5.2   Routing feedback - adaptive SCA

In the previous sections, the evaluation of the SCA technique did not entail any feedback from the routing protocol. It is possible to determine an optimal SCA parameter for a given hop count and apply the optimised parameter to the TCP agent for any given path length. To determine such a parameter per hop count we utilise the same methodology as is done in [25] for the adaptive CWL method. Unless otherwise noted, the simulation parameters are as set in the previous section.

Specifically, string topologies of various lengths are set up ranging from 4-15 hops. A TCP Reno connection carrying FTP traffic is established between the end-nodes.

The FTP source is active from the beginning of the simulation until its end at the 900 seconds mark. The SCA strategy is applied by varying the SCA parameter in the range of 0 to 50. The value of the threshold parameter that yields the best goodput in each string topology is deemed to be efficient and is noted. Table 4.4 shows which parameter was optimal for each hop count when using the DSR protocol. We have also conducted the same experiments using the AODV protocol, which reveals that the optimal parameters per hop count are very similar to the ones for DSR. The relevant table is included in Appendix B.2.

The idea of an adaptive SCA strategy is similar in principle to the adaptive CWL method [26]. Using the route path length information provided by DSR, TCP can dynamically adjust the SCA parameter whenever the length of the route changes. As DSR is a source routing protocol, a cooperating TCP agent is guaranteed to be aware of the total hop count to the destination (as it sets the path in the packet header).

We have evaluated this by performing simulation experiments in the following fashion. In a flat space of 1500x300m, 50 nodes were placed randomly. The random waypoint mobility model is used to simulate mobility by assuming continuous mobility (0 seconds pause time) and maximum node speeds of 2, 5 and 10m/sec. For each pause time/maximum node speed combination 50 different topologies are created and each simulation lasted for 900 seconds. The metric collected at the end of the simulation run was goodput.

Results from these experiments are included in Figure 4.21 for both static and adaptive SCA agents. The benefits of an adaptive SCA strategy were marginal at best (the difference between static and adaptive ranging from -1% to 1.5%), and not significant mainly because there was not much difference between the various SCA rates in terms of goodput in static topologies as evident in Figure 4.6. A much slower sending rate increase in the congestion avoidance phase can be unproductive in the case of frequent route breakages, though, because the TCP source cannot utilise the full capacity of the link in the little time it is available (i.e. before route breakages occur). The effects are especially noticeable if hidden terminal effects cannot possibly take place (i.e. the path is less than 4 hops long). At least in single flow environments, the adaptive SCA strategy does not appear to be beneficial over static SCA, however as noted before, it is possible to use route length feedback to deactivate the SCA mechanism when hidden terminals are not an issue (i.e. in this case, when the route length is less than 4 hops).

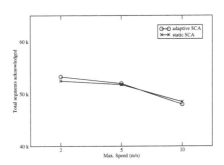

Figure 4.21. Goodput against node speed for the adaptive and static SCA Reno

Table 4.4. Default SCA Reno parameter on string topologies for DSR

Hop Count(h)	parameter	h	parameter
4	42	10	48
5	45	11	35
6	39	12	10
7	49	13	50
8	47	14	25
9	29	15	10

Finally, note that although simulation results in the case of other pause time periods are not included, these would not be expected to denote different behaviour; the factor which precludes significant performance difference between static and adaptive SCA (i.e. that an SCA parameter of 10 is a "good enough" value) would still hold.

4.6   Conclusions

Inspired by the conservative sending rate increase of TCP Vegas, and motivated by its compelling performance advantage over Reno-based TCP variants, this chapter has introduced a new mechanism, named Slow Congestion Avoidance (SCA), which employs a more conservative sending rate increase and which alleviates some of the intra-flow spatial contention caused by traditional TCP agents. To this effect, the new method employs a parametrised delay in the growth of the TCP congestion window, which is implementable in both the slow start and congestion avoidance phases of Reno-based variants. This work has examined the possible applications of the new technique and has shown, with the aid of detailed simulation traces, that it is most effective when applied to the congestion avoidance phase of TCP as this is mostly in effect in long-lived TCP flows. The resulting technique has been named SCA TCP and is orthogonal to link layer solutions to spatial contention as it is end-to-end applicable and involves only transport layer alterations.

The technique has been shown to improve goodput by 4-26% in the case of TCP Reno in a variety of dynamic topologies matching and even surpassing TCP Vegas' performance without incurring the latter's computational overhead. Further, the new technique has been contrasted with an existing solution towards spatial contention alleviation, namely the adaptive Congestion Window Limit (CWL) method. Both methods were employed in dynamic *ad hoc* topologies, using long-lived TCP flows for both AODV and DSR protocols. The subsequent evaluation has revealed SCA to outperform adaptive CWL in terms of goodput under various mobility conditions by 4-20%.

Since the SCA technique has been shown to result in goodput gains by alleviating intra-flow spatial contention, subsequent work in this chapter has also investigated its applicability to multiple TCP flows originating from the same sender. The SCA method has been further parametrised to deal with inter-flow spatial contention in the case of multiple TCP flows and was shown to outperform the plain TCP agent in the case of NewReno in dynamic topologies. Specifically, the performance improvement achieved in various dynamic topologies exhibiting various degrees of mobility was 4-12%.

The possibility of utilising feedback on the path length as realised by the routing protocol has also been examined with the prospect of customising the SCA parameter on a per hop-count basis. However, and in the case of the DSR protocol which provides such feedback, it has been shown that the utilisation of such a technique does not lead to significant improvement gains as the default SCA parameter provides an equivalent goodput improvement.

In this chapter, the SCA technique has addressed the issue of alleviating spatial contention via modifications on the sending side in a communicating pair. However, spatial contention is also caused by the receiving entity in a TCP communicating tuple, which returns feedback to the sender through the injection of acknowledgement (ACK) segments in the network. The next chapter reviews the literature on the topic of reducing acknowledgement traffic caused by the receiver (termed "ACK-thinning" in research nomenclature [6]) and identifies previously ignored problems with existing proposals. Then, a combination of MAC layer options in 802.11 transceivers is evaluated in tandem with existing ACK-thinning techniques to result in goodput improvements.

Chapter 5

ACK-thinning techniques in MANETs

5.1 Introduction

As identified in previous work in MANETs [26, 43, 104, 105, 107], the issue of spatial contention in multihop wireless networks, which is aggravated by the existence of hidden terminals, is caused by the inability of the MAC mechanism to properly coordinate transmissions. Specifically, for unoptimised TCP agents, too many segments may be injected into the pipe at any one time and the MAC mechanism may be unable to handle those numerous elements competing for transmission time. As discussed in the previous chapter, spatial contention could be mitigated through alterations at the transport layer by applying changes to the TCP sender. Although changes at that level may not eliminate the problem in its entirety, it is, nonetheless, possible to enhance throughput especially across long transmission paths.

In the context of a TCP communicating pair there are two elements contributing to spatial contention; the sender's transmission of TCP DATA segments and the receiver's reciprocal ACK response. The SCA approach outlined in the previous chapter aims to reduce the amount of outstanding DATA segments, i.e. segments injected by the *sender*. By considering the complementary aspect of the problem, it would therefore be beneficial to reduce the amount of ACK traffic, as generated by the *receiver*. Mechanisms to that effect are named *ACK-thinning* techniques. An optimisation of this nature has already been implemented in the TCP standard [54] as, notably, TCP incorporates a piggyback mechanism for ACK segments by including the ACK byte

117

sequence number in DATA segments exchanged between hosts. Intuitively, in the case where the DATA traffic derives mostly from one of the two communicating parties such a measure is ineffective as it does not come into effect frequently. Hence, it can be beneficial to introduce other, possibly orthogonal, end-to-end techniques to deal with the issue.

Specifically, several previous research studies [6, 26, 31, 112] have identified the receiver's ACK response as an important cause of spatial contention and have targeted its reduction, either by piggybacking TCP ACKs on other traffic along the same path [112] or by reducing the ACK response frequency [6, 31]. Solutions present in the literature also include the activation of the optional delayed ACK TCP mechanism [105, 107] and an appraisal of its effectiveness. However, these studies have only performed limited evaluation of their proposed changes in MANET topologies and have largely ignored limitations imposed by real-life TCP implementations such as the overhead of the TCP timer granularity. In particular, in several research works [6, 26, 31] it is assumed that the TCP ACK response may be delayed with exact precision, whilst in modern TCP implementations, delaying an ACK response entails some degree of granularity. Moreover, there has been little investigation into quantifying the level of spatial contention added by TCP ACK responses as opposed to contention contributed by DATA segments. Such an enquiry could provide insight on the degree of improvement that may be achievable with ACK-thinning methods. Finally, although the RTS/CTS exchange of the 802.11 protocol has been identified as a source of spatial contention [103], the MAC mechanism configuration as used in the literature with respect to spatial contention issues [6, 26, 31, 43, 71, 85, 104, 105, 108], does not include considerations of optimisations possible within the 802.11 specification.

After having identified shortcomings in the evaluation conducted by previous research with regard to ACK-thinning mechanisms, our first contribution in this chapter is a discussion regarding the effects of those limitations on the results derived in past works. Significantly, it is highlighted that previous attempts at reducing spatial contention by affecting ACK responses have largely ignored inherent limitations of existing TCP implementations, which, in turn, make the proposed changes difficult to implement. As such, it has been assumed in past research [6, 31] that ACK responses may be delayed for any finely defined time period, disregarding the delay timer granularity. Further, concerns may be raised with respect to the path length characteristics assumed in existing literature [6, 26, 31], where large hop counts are emphasised in importance, but which are unlikely given a reasonable network size [60, 92] and appear rarely in popular research mobility models, like the random waypoint model [110]. Further, two 802.11 compatible MAC layer optimisations with respect to the RTS/CTS exchange are introduced. The first is applicable to the ACK-responses of the TCP agent, whilst the second applies to both ACK and DATA frames. Both are shown to result in a reduction of spatial contention and to have a positive impact on TCP goodput.

The rest of the chapter is organised as follows. Section 5.2 describes two existing end-to-end ACK-thinning techniques (delayed ACKs and Dynamic Adaptive ACKs) and highlights their degree of efficiency as outlined in the literature. Then, Section 5.3 identifies problems and limitations on the evaluation of the previously outlined solutions. Three problem areas are identified, namely the requirements in temporal granularity of ACK responses, identifying which path length cases are more likely to occur and thus be targeted for optimisation, and the effect of making use of the RTS/CTS mechanism of the 802.11 protocol. Section 5.4 presents a possible MAC layer optimisation on 802.11 compliant devices, which has been largely ignored in existing literature, and notes its positive result on goodput. Section 5.5 analyses the performance gains of combining this optimisation with the implementable end-to-end ACK-thinning mechanisms introduced previously in a variety of topologies. Finally, Section 5.6 concludes the chapter and offers an overview of the presented results.

5.2   ACK-thinning in MANETs

This section presents two ACK-thinning mechanisms introduced previously [31, 105]. The first one, delayed acknowledgements (ACKs), is an optimisation strategy which has been developed for wired networks but which has also been shown to have significant merits in a MANET setting [105]. The second optimisation is named Dynamic Adaptive Acknowledgements and is intended for use in MANETs as an efficient ACK-thinning strategy [31], being applicable to both long-lived and short-lived flows. This section, thus, offers prerequisite background information on ACK-thinning methods which supports the discussion for the rest of this chapter.

5.2.1 Delayed acknowledgements

The delayed acknowledgements mechanism is an oft-enabled [4] feature of TCP, as first described in [27]. Its principal operation is simple and relies on the cumulative nature of TCP acknowledgements; instead of immediately replying with an ACK upon receiving a DATA segment, the TCP receiver waits a short time (usually 100-500ms [3]). If a subsequent DATA segment arrives, assuming it is consecutive and in-order, then it is possible to inject a single ACK into the pipe, which cumulatively verifies the receipt of both DATA segments. Further, since TCP connections are duplex, it is also possible to piggyback the ACK onto DATA segments being sent in the other direction, i.e. from the 'receiver' to the 'sender', thus saving bandwidth. To avoid confusing TCP estimates, such as the round trip time estimator and TCP's ACK-clocking mechanism, the relevant RFC [15] dictates that ACKs should not be delayed in any case for more than a single (extra) DATA segment, or for a time period of more than 500ms.

It has been shown in the literature that delayed ACKs are beneficial for TCP throughput and should be enabled by default [3]. Subsequent research in wired networks, has investigated the possibility of increasing the delay response for TCP so as to alleviate the competition of ACKs for bandwidth space along with TCP DATA. However, it has been demonstrated that this might result in "burstiness" or has other shortcomings, especially with regard to wide deployment on the Internet [90]. In MANETs, the throughput enhancing property of delayed ACKs has been demonstrated repeatedly in static and dynamic topologies, with reactive [106] and proactive [86] TCP agents. In special cases the improvement in TCP throughput is in the range of 15-32% [105].

5.2.2 Dynamic Adaptive Acknowledgements

The Dynamic Adaptive Acknowledgement (DAA) method is a sender/receiver modification introduced by d' Oliveira et al. [31]. It aims to reduce the number of ACKs produced at the receiver by taking advantage of their cumulative property. The DAA method dictates changes to the TCP sender as well as the receiver with the delay of ACK responses being performed in a dynamic manner so as to adapt to changing network conditions. There is some processing overhead associated with DAA but the trade-off is a general increase in throughput and better utilisation of the wireless channel (improving on spatial reuse).

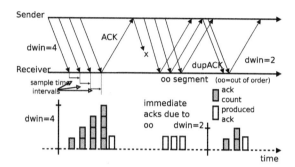

Figure 5.1. Demonstration of Dynamic Adaptive ACKs

The main operation of DAA is based around a mechanism of withholding ACK responses at the receiver. The only requirement at the sender is to restrict its congestion window (*cwnd*) between 2 and 4 segments, i.e. allow it to keep 2,3 or 4 segments outstanding in the network at any one time. The receiver maintains a dynamic delaying window (*dwin*) with size ranging from 2 to 4 full sized segments, which determines when an ACK will be produced. Whenever a consecutive DATA segment is received, an *ack_count* variable increases by one until it reaches the current value of *dwin*. When *ack_count* = *dwin* an ACK response is immediately produced, *ack_count* is reset to one segment and the value of *dwin* increases by one. This signifies the beginning of the next *epoch*, that is to say the next group of DATA segments for which the corresponding ACKs will be delayed. Note that the *ack_count* variable differentiates between these *epochs* and is initially set to one segment.

If every DATA (and ACK) segment is successfully delivered the DAA method allows eventually 4 DATA segments to produce one ACK response from the receiver. However, as DATA segments may be lost or be overly delayed during transit it is useful to introduce a mechanism whereby a prompt ACK response could still be triggered without waiting for *ack_count* to reach the current *dwin*. The proposed prompt response mechanism in DAA works as follows. For each DATA segment received, say $i, i + 1, i + 2, \ldots$, and for which an ACK is to be delayed, its inter-arrival time gap with the previous DATA reception is recorded, say $\delta_i, \delta_{i+1}, \delta_{i+2}, \ldots$). Effectively, for each ACK delay *epoch*, the inter-arrival times of incoming DATA segments are noted.

These collected time periods are used to calculate a smoothed average which signifies an "expected" inter-arrival time, say $\bar{\delta}_{i+1}$, for consecutive ACK segments. The calculation is performed using a low-pass filter and is used to assess a timeout interval for the ACK response. If $\bar{\delta}_i$ is the last average calculated, δ_{i+1} is the DATA segment inter-arrival time sampled and α is an inter-arrival smoothing factor, with $0 < \alpha < 1$, then

$$\bar{\delta}_{i+1} = \alpha * \bar{\delta}_i + (1 - \alpha) * \delta_{i+1} \qquad (5.1)$$

As the relevant RFC suggests [5], in the case of out-of-order segments an ACK response is immediately prompted, but otherwise the receiver waits for a time period T_i before responding. This effective timeout interval is calculated with a timeout tolerance factor, κ, with $\kappa > 0$ as shown in equation (5.2), where $\bar{\delta}_i$ is calculated by equation (5.1).

$$T_i = (2 + \kappa) * \bar{\delta}_i \qquad (5.2)$$

Note that for short file transfers it may be desirable to produce quick ACK responses so as to allow a increase of the sending rate during the slow start phase at the sender. To this end, there exists a mechanism in DAA method to account for variable, as opposed to fixed, increases in the *dwin* size. There is a speed increase factor, μ, with $0 < \mu < 1$. If *maxdwin* is a status indicator which turns true when the maximum possible value for *dwin* has been reached (by default set to 4 segments) then *dwin* growth is set to

$$dwin = \begin{cases} dwin + \mu & \text{if } maxdwin\text{=false,} \\ dwin + 1 & \text{otherwise.} \end{cases} \qquad (5.3)$$

Equation (5.3) allows the receiver to respond immediately with ACKs in the case when the TCP sender is in the slow start phase where each ACK increases *cwnd* by a single segment. If ACKs were delayed during this phase the sender would not receive enough ACKs to increase its sending rate effectively. Essentially, the *maxdwin* parameter signifies (at the receiver) when the slow start phase (at the sender) is over. Once the *maxdwin* is reached once, then this mechanism is not activated again for the same connection. Hence, this facility is intended for short file transfers.

The DAA method has been evaluated on string and mesh topologies of varying length and different number of flows [31]. On string topologies of up to 8 hops and 20 flows the method increases throughput up to 50% over plain TCP NewReno. In particular it is noted that as the number of concurrent flows increases, the DAA method becomes increasingly effective. On mesh topologies, taking into account 3 and 6 cross traffic sources the performance improvement is of the same magnitude, but the difference diminishes against optimised (with respect to maximum *congestion window* size) Vegas and SACK agents. The method does not lead to a discernible performance advantage in the case of short-lived flows.

From an implementation perspective the DAA method is an end-to-end sender/receiver modification. The sender needs to be tweaked with respect to initial *cwnd* size and the receiver needs to implement the ACK dynamic window. However, no modifications are required to other layers and no cross-layer feedback is assumed.

5.3 Existing evaluation of ACK-thinning techniques

This section outlines three factors with respect to the evaluation of ACK-thinning techniques that have either not been considered in their entirety in the existing literature or have been ignored altogether [71]. In particular, the requirements of the TCP delayed ACKs timer granularity [31], considerations of the hop-count in general simulation topologies [43, 105] and issues with the implementation of the 802.11 RTS/CTS exchange in existing evaluation studies [19, 43, 71, 105] have only partially been addressed. This section, thus, presents arguments for a more complete approach in evaluating ACK-thinning techniques and forwards recommendations for a more comprehensive evaluation.

Simulation Setup: The simulations performed in this section all share the following parameters. The signal propagation model used is the Two-Ray Ground model, and the wireless transceivers are modelled after the Lucent WaveLan II models [64] (as done throughout this dissertation). The TCP agent used is NewReno and the segment size is set to 1460 bytes. The routing protocol used is AODV. A more complete list of parameters for both the routing and transport agents is included in Appendix A, and both mirror the setup used in previous chapters. Finally, the string topology simulations involve topologies arranged in the manner depicted in Figure 5.2 and thus, are set up as in the previous chapters in this dissertation.

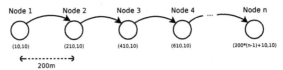

Figure 5.2. n-node string topology

5.3.1   Granularity of the TCP timer

Past research has focused on introducing delay when injecting ACK segments, as a means of bundling consecutive ACKs into fewer batches by taking advantage of their cumulative property [6, 31, 105]. The amount of delay introduced in such cases is variable and the techniques introducing it may be computationally intensive. However, there is scarcely a mention on how fine the control over such a delay need be [31, 51, 70].

Specifically, it is assumed that upon deciding on a given delay, the receiver will be able to delay the transmission of the ACK segment for precisely that amount of time [31]. However, actual TCP implementations may not operate in such a fashion, but instead implement timers with certain granularity. Previous research on TCP implementations has identified issues arising from using a coarse "heartbeat" timer for TCP agents, especially with respect to the RTO mechanism [78, 89]. In general, the relevant RFC [15] demands an ACK response to be no later than 500ms, which is a rule compliant Internet hosts must adhere to and which is in fact implemented as standard in popular operating systems. Specifically, the granularity implemented is of the order of 10ms (Linux kernel 2.6 [76]), 100ms (FreeBSD v5.0 [11]), or even 500 ms [15]. Research dealing with ACK-thinning methods has not taken such a requirement into account and has instead opted for very fine (or even infinitely fine) ACK timers [31, 51, 70].

It may not be claimed that the granularity specified in the systems above is going to be a typical quantity in TCP agents deployed on MANET nodes. It should be stated

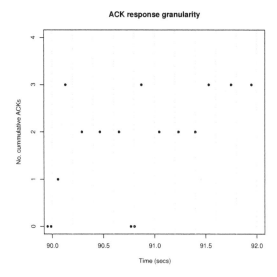

Figure 5.3. Illustrating "heartbeat" timer granularity

nonetheless, that even if MANET agents implement sufficiently fine-grained timers so that their precision is not a performance altering factor, these would still have to interact with Internet hosts, which would not necessarily conform to such fine-grained timer requirements. As such, the issue is worthy of some research effort to determine how timer granularity may affect an existing implementation.

To further explore the last point we have implemented the DAA method as outlined in Section 5.2.2 in the ns-2 simulator [37]. An FTP transfer session with an infinite backlog is initiated at the beginning of the simulation between the two endpoints of a 7-hop string topology and lasts for the duration of the simulation, which is 120 secs. Figure 5.3 depicts the ACK responses from the receiver during the 90-92 secs interval. Each filled point indicates the number of ACK responses delayed before launching a cumulative ACK. A value of 0 denotes an immediate ACK, as a response to out-of-order segments. The dotted vertical lines indicate the ACK checkpoints of a

Table 5.1. Effects of timer granularity on the DAA technique

Hops	Granularity				
	0ms	10ms	50ms	200ms	500ms
3	6237	6299(1%)	6311(1%)	6305(1%)	6293(-0.8%)
4	4370	4388(0.4%)	4295(-1.7%)	3900(-12%)	3504(-24.7%)
5	3777	3727(-1.3%)	3621(-4.3%)	3517(-7.3%)	3011(-25.4%)
6	3257	3320(1.8%)	3231(-0.8%)	3043(-7%)	2618(-24.4%)
7	3129	3064(-3%)	2997(-4.4%)	2851(-9.7%)	2450(-27.7%)
8	2910	2873(-1.2%)	2797(-4%)	2686(-8.3%)	2320(-25.4%)
9	2800	2747(-1.9%)	2679(-4.5%)	2516(-11.2%)	2156(-29.8%)
10	2652	2624(-1%)	2542(-4.3%)	2429(-9.1%)	1962(-35.1%)
11	2465	2474(0.3%)	2517(-2%)	2459(-0.2%)	1963(-25.5%)

"heartbeat" timer with 200ms granularity. If simulation were to allow for such granularity, the ACK responses would only occur on each "heartbeat", i.e. at the time points
indicated by the vertical lines. However, in this case since such provisions are not
made, ACK responses occur at any point in time (infinite timer granularity).

To quantify the effects of variable timer granularity, the DAA method was deployed
on string topologies of variable length n, with $4 \leq n \leq 12$. An end-to-end FTP transfer with infinite backlog was set up as before for 120 secs. At the end of the simulation
the achieved TCP goodput was noted. Each simulation was repeated by setting the
"heartbeat" granularity to 0 (immediate response),10, 50, 200 and 500 ms. The values
were chosen to be representative of the ideal case (0ms), the case of existing operating
systems (10,50,200) and the case of older systems (500ms - BSD 4.4 [18]). The effects
of different granularity per hop-count on goodput can be examined in Table 5.1 where
goodput as the total number of segments transmitted, without considering retransmissions, is indicated. The percentage in parenthesis indicates the difference in goodput
for each granularity level as compared to the ideal case (0ms granularity). Note that
as noted in the previous chapters, as the hop count increases, goodput decreases as a
side-effect of spatial contention [26, 106].

In general, there is little difference in performance between the ideal case of 0ms granularity and those of 10 and 50ms for any hop count. For a receiver with a 200 ms level of granularity there is a penalty in goodput of approximately 7-12% for string topologies of length 5-11 nodes (4-10 hops). In the case of 500ms granularity the goodput performance is consistently worse by approximately 25%. Evidently, in the case of the string topology and especially over long paths, goodput suffers if the path is sufficiently long (over 5 hops) and the receiver ACK granularity is over 200ms. In order to be representative of a modern operating system the rest of the chapter assumes a Linux kernel-like 10ms granularity for the receiver. Contrary to previous research efforts [6, 31] this work explicitly states that assumption.

Although the case of the string topology is a special case of a communications path, it can be representative of long lived path behaviour in a dynamic topological scenario and is, thus, instructive on how granularity requirements may effect goodput. The effects of granularity in this case are discussed with respect to the DAA method in particular but apply to any other method requiring fine-grained controlled delay in ACK responses [51, 70].

5.3.2   Path length

Studies on ACK-thinning in MANETs have largely been conducted in the special case of the string topology [6, 26, 43, 105, 107] as this static setup offers a convenient environment to study the effects of spatial contention in isolation from from the effects of mobility. More recent research efforts [31, 43] have also focused on analysing ACK reductions in the context of mesh topologies as these offer the benefit of studying spatial contention caused by several flows operating in tandem.

The random waypoint model has been a popular mobility pattern template, used widely in previous MANET research [19, 30, 34, 71, 104]. Although other mobility models exist [20], it has been deemed generic enough to at least warrant some consensus as the standard parameter in an evaluation setup [75]. Since string topologies are the testbed for the evaluation of ACK-thinning techniques it becomes interesting to investigate an average value of the formed paths in topologies generated by the random waypoint model. Such an endeavour would illustrate which range of length for the string topology is of most interest assuming that a given path length range most frequently noted in random waypoint generated topologies is also one that would likely be encountered in real-life deployments.

To investigate this aspect, we have conducted a set of path measuring experiments in a variety of dynamic topologies, set in motion according to the random waypoint model under low (2m/s), medium (5m/s) and high (15m/s) mobility conditions. The simulation time was set to 900 secs, as in the experiments considered in Chapter 3 and as widely practised in literature [25, 34, 49, 87]. For each mobility condition, 200 topologies consisting of 50 nodes each were taken into account, occupying both square (1000x1000m) and strip (1500x300) areas. The transceiver and signal propagation model used is the same as in the string topology simulations above.

The path length measurement was conducted in the following fashion. The average path length $p_{i,j}$ between nodes i and j was set as an average of the path length (distance in hops) between nodes i and j during simulation time when such a path did exist; in the case of a disconnection (as long as that might have been) nothing was added or subtracted from the average. It should be indicated that $p_{i,j} = p_{j,i}, \forall i, j$ as the path length between two nodes is the same regardless of which end-node is considered as the beginning point for the hop-count. Further note that the shortest possible path in terms of hop distance was considered as the actual path length between nodes i and j at any one time. This is considered a fair representation of the actual path length, as all the routing protocols under examination in this dissertation, namely AODV [92], DSR [60] and OLSR [48], contain mechanisms to favour the shortest possible path for communications. However, there are cases where the shortest path may not be discovered by the routing protocol, for instance due to the broadcasting storm problem [100] or otherwise severe spatial contention which might lead to some route request drops, but we assume these cases to be relatively rare and expect our findings to hold true in the case when an actual routing agent is used.

Figure 5.4 shows histograms for strip (1500x300m) topologies under low (2m/s), medium (5m/s) and high (15m/s) mobility. It is evident that the path length rarely exceeds 3 hops under these conditions. Figure 5.5 depicts the relevant histograms for square topologies (1000x1000m), where the same conclusions may be extrapolated. Note that the breaks between the histogram cells (bins) were set according to the Freedman-Diaconis rule [55]. Appendix B.1 also contains, for completeness, histograms produced with the Sturge's rule, which lead to the same conclusion.

Overall, there is strong indication that short communicating paths are important

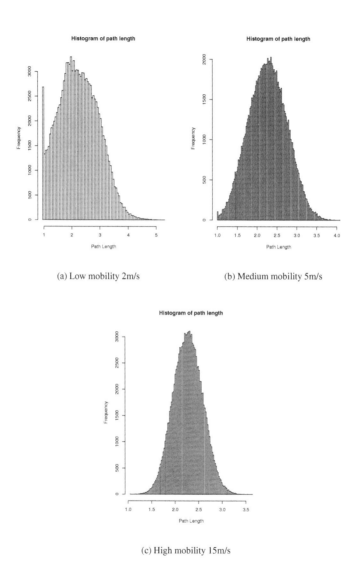

(a) Low mobility 2m/s   (b) Medium mobility 5m/s

(c) High mobility 15m/s

Figure 5.4. Histogram of path lengths in a strip area (1500x300m)

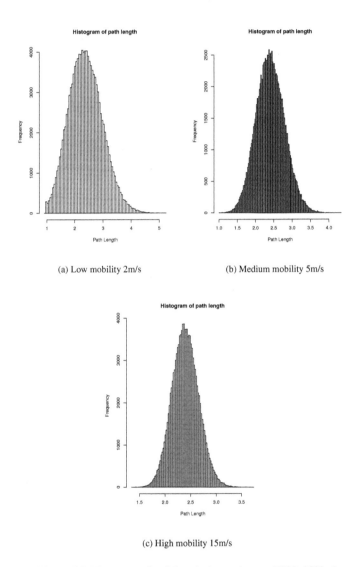

(a) Low mobility 2m/s   (b) Medium mobility 5m/s

(c) High mobility 15m/s

Figure 5.5. Histogram of path lengths in a strip area (1000x1000m)

enough to be emphasised during evaluation and simulation tracing experiments. For the rest of this chapter and in subsequent evaluation, our focus is equally distributed over long and short paths without bias, in contrast to previous research [6, 26].

5.3.3 RTS/CTS exchange

The Request to Send/Clear to Send (RTS/CTS) short frame exchange is defined by the 802.11 specification [52] as a means to alleviate the hidden and exposed terminal effects, described in Section 1.1.1.

The principle operation of the RTS/CTS exchange is as follows; before transmitting a DATA segment to a particular neighbouring node, the sender first transmits an RTS frame which acts as an intent-to-transmit message containing the destination and duration of the intended transmission. The intended receiver, then, responds in turn with a CTS frame which includes similar timing information, informing neighbouring nodes of the length of the subsequent intended transmission. The procedure is depicted in Figure 5.6. Specifically, the dotted lines in Figure 5.6(b) illustrate that a non-intended receiver of the DATA frame transmitted by the sender, simply defers transmission until the medium is perceived to be idle for sometime. Following the deferral, the node in question may again resume contending for transmission time. In the case where RTS/CTS is used (Figure 5.6(b)) the transmission time for the same frame size increases as there is the added overhead of the RTS/CTS frame exchange in addition to the DATA and ACK frames. Note that in this case, however, nodes within both the transmission radius of the sender and the receiver are informed of the impending DATA frame exchange so as to withhold their own transmissions. This is in contrast to the previous case where the mechanism is not used, and only nodes within the sender's communications range are aware to refrain from transmitting at the same time and, thus, avoid causing a collision.

As stated, the extra overhead of the RTS/CTS mechanism cannot always be justified for every DATA frame transmission. This is acknowledged in the 802.11 specification [52], especially with regards to short DATA frames. As such the 802.11 standard implements RTS/CTS control through the *dot11RTSThreshold* attribute. This allows the use of RTS/CTS to be active for all frames, frames longer than a specified length or not at all. In particular, if the number of bytes in the segment to be transmitted

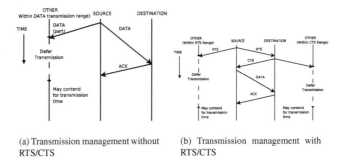

(a) Transmission management without
RTS/CTS

(b) Transmission management with
RTS/CTS

Figure 5.6. Illustration of wireless communication with and without the RTS/CTS exchange

is below the *dot11RTSThreshold*, then the RTS/CTS exchange is not performed. The default value in the 802.11 standard is 2347 bytes, which implies disabling the exchange altogether [52] (as it is typically not necessary in infrastructure wireless LANs). Nonetheless, it is useful to activate the feature when all frames need to be "protected" against the hidden terminal effect or even activate it selectively for large frames only whilst leaving it turned off for shorter frames (say TCP ACK segments), as these may be less at risk as they can be transmitted faster [103].

Previous research on ACK-thinning techniques has not taken into account the effects of the RTS/CTS mechanism, largely assuming that the mechanism would be active at all times [6, 26, 31]. However, for 802.11 conforming implementations, the RTS/CTS mechanism would be disabled by default for TCP DATA and ACK frames, since the encapsulating frame would not be of sufficient size (assuming that the *dot11RTSThreshold* was set to the default value). In this chapter we investigate the implications of disabling the RTS/CTS exchange and note the throughput performance implications. Disabling RTS/CTS may be done in the case of only ACK segments (using the *dot11RTSThreshold* parameter) or for both TCP DATA and ACK segments (by disabling RTS/CTS altogether). Either method may be used in tandem with ACK-thinning techniques and possibly lead to a cumulative performance improvement.

5.4   Performance impact of RTS/CTS

This section demonstrates with the aid of a simulation trace example the effect of disabling the RTS/CTS response for ACK segments, as well as for both ACK and DATA
TCP segments. To this end, two new metrics on spatial contention are introduced and
are further used in the next section for general string and mesh topology evaluation.

The experimental setup is as follows. A string topology of 5 nodes (4 hops) is
assumed in the fashion depicted in Figure 5.2, and as used in the previous section. An
FTP session is initiated at the beginning of the simulation between the two end-points
of the string topology and continues for 120 secs at which point the simulation ends. As
in the previous section, the TCP agent used to carry the FTP traffic is NewReno [38].
The rest of the TCP parameters are as noted in Appendix A. Note that although in
this particular case delayed ACKs are not employed (which is a common optimisation
enabled by default in some TCP implementations [4]), the observations made are applicable even in the case when such a mechanism is opted for. Section 5.5 contains
more discussion on this point and includes further results in the case of string and mesh
topologies.

To *quantify* the effects on spatial contention of the RTS/CTS exchange two metrics
are introduced. The first metric is the number of DATA frames dropped due to *repeated
failed* MAC layer retransmissions. Note that the maximum number of MAC layer re-
tries for a frame is set to 4 attempts as per the 802.11 specification[1] [52]. The payload
of these frames is either a TCP DATA or ACK segment and so a series of repeated
transmission failures leading to a drop is marked as either $FAIL_{DATA}$ or $FAIL_{ACK}$ respectively. The second metric is the number of failed RTS/CTS exchange *procedures*.
It is worth mentioning that an RTS/CTS exchange is attempted several times by the
MAC mechanism before it is marked as having failed. The required number of such attempts is 7 in the 802.11 specification[2]. Such failed attempts are noted as $FAIL_{RTS/CTS}$
drops. Further, the number of collisions is noted during the FTP transfer. Such collisions may be MAC frames containing TCP DATA, ACK or RTS/CTS payloads and
so are marked COL_{DATA}, COL_{ACK} and $COL_{RTS/CTS}$. It should be specified that a
high number of collisions indicate an *increasing degree* of spatial contention, whilst a

---

[1]In the 802.11 specification the parameter is named *dot11LongRetryLimit*
[2]In the 802.11 specification the parameter is named *dot11ShortRetryLimit*

high number of failed negotiations, either in TCP DATA or ACK transfers, denotes an *increasing inability* of the distributed MAC mechanism (the Distributed Coordination Function in 802.11 nomenclature [52]) to effectively cope with spatial contention.

The simulation is run three times and with each iteration a different RTS/CTS strategy is employed. In the first round the RTS/CTS exchange is fully utilised in both TCP DATA and ACK segment exchanges. Subsequently, the RTS/CTS mechanism is only opted for "sufficiently large" TCP segments, i.e. only for DATA segments. Finally, in the third iteration, the RTS/CTS exchange is eliminated altogether. These three strategies are hereafter referred to as "Full RTS/CTS", "Partial RTS/CTS" and "No RTS/CTS" respectively. The discussion that follows is conducted with the aid of measurements of spatial contention expressed with the metrics introduced in this section.

Figure 5.7 depicts a 101-running average of the number of segments in flight throughout the simulation for the three different strategies. The graphs depict the number of DATA and ACK segments existing along the path through the simulation time and also shows their combined (aggregate) presence. A visual inspection of the figures reveals that disabling RTS/CTS altogether (Figure 5.7(c)) results in the TCP agent being able to maintain more segments in the pipe at any one time in both its receiving and sending aspects, i.e. both for DATA and ACK segments. In this particular case, on average, 15.74 segments exist in the pipe at any one time using the "No RTS/CTS" strategy which is significantly higher (by 183.76% and 167.4%) than the averages of 5.548 and 5.886 segments achieved by the "Full RTS/CTS" and "Partial RTS/CTS" strategies, respectively. The complete numerical set of averages for all three strategies, categorised by type (DATA or ACK or both) is shown in Table 5.3. In this case, as for the rest of this section, table entries may be accompanied (where applicable) with a number in parenthesis denoting the numerical difference (percentage-wise) between the value examined for that particular strategy against the value achieved under the "Full RTS/CTS" strategy.

Overall, the "No RTS/CTS" strategy allows the MAC mechanism to be more efficient in coordinating the transmissions of a higher number of outstanding TCP segments. Table 5.2 contains the number of collisions and overall transmission failures for

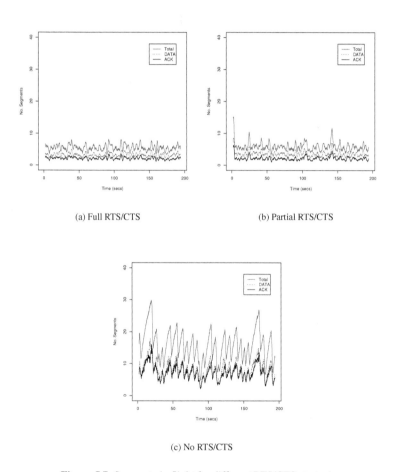

(a) Full RTS/CTS   (b) Partial RTS/CTS

(c) No RTS/CTS

Figure 5.7. Segments in flight for different RTS/CTS strategies

Table 5.2. Frame collisions and drops for each RTS/CTS strategy

	Full RTS/CTS	Partial RTS/CTS	No RTS/CTS
$\text{COL}_{RTS/CTS}$	15047	8236 (-45%)	0 (-)
COL_{DATA}	79	240	3696
COL_{ACK}	67	6318	8197
COL_{TOTAL}	15193	14794	11893
$\text{FAIL}_{RTS/CTS}$	273	169 (-38%)	0 (-)
FAIL_{DATA}	167	169 (1.1%)	35 (-79.04%)
FAIL_{ACK}	106	81 (-23.5%)	310 (192%)
FAIL_{TOTAL}	546	419 (-23.2%)	345 (-36.8%)

each RTS/CTS strategy. Note that no segment drops were recorded due to buffer over-flows in the forwarding nodes. Further, Table 5.4 contains the goodput achieved (in to-tal segments transmitted and ACKed) for each strategy. Both the "Partial" and the "No RTS/CTS" strategies suffer from fewer frame collisions overall, but incur greater TCP and ACK frame collisions than the "Full RTS/CTS" method. Similarly, a greater num-ber of RTS/CTS failed negotiations occur in the case of the "Full RTS/CTS" method as compared to the "Partial RTS/CTS" strategy ("No RTS/CTS" does not employ this exchange and thus records no such failures). This provides some indication of the over-head added spatial contention provided by the RTS/CTS exchange; collisions among segments increase and the MAC mechanism is unable to effectively coordinate trans-missions (as denoted by the average number of TCP segments maintained in the pipe).

It is evident that the increased number of RTS/CTS transmissions leads to several RTS/CTS collisions which have a detrimental effect on goodput. An indication of this is the number of failed TCP and ACK transmissions for the "Full RTS/CTS" method. The number of TCP DATA and ACK segment drops cannot be solely attributed to MAC frame drops containing DATA or ACK payloads. The number of collisions of those are too few (79 and 67 segments respectively) to account for the number of frame drops (167 for TCP and 106 for ACK payloads). Hence the increased number of RTS/CTS exchange failures as compared to the other methods (273 in the case of "Full RTS/CTS" as opposed to 169 for the "Partial RTS/CTS" method) as well as the number

Table 5.3. Average segments in flight for each RTS/CTS strategy

	Full RTS/CTS	Partial RTS/CTS	No RTS/CTS
Mean$_{TCP}$	3.278	3.602 (9.8%)	8.337 (154.3%)
Mean$_{ACK}$	2.148	2.156 (0.3%)	7.313 (240.4%)
Mean$_{TOTAL}$	5.548	5.886 (6%)	15.74 (183.7%)

Table 5.4. Goodput achieved for each RTS/CTS strategy

	Full RTS/CTS	Partial RTS/CTS	No RTS/CTS
TCP Goodput (in total no. of segments)	4476 (-)	4748 (6%)	5365 (19.8%)

of RTS/CTS collisions (45% greater than the "Partial" strategy) can largely account for the discrepancy in goodput (where the Full RTS/CTS method incurs performance hit of 6 and 19.8% compared to the Partial and No RTS/CTS techniques respectively). Note that an RTS/CTS exchange failure (that is 7 consecutive failed attempts) results in a TCP DATA or ACK segment drop corresponding to that transmission process. This fact explains why there are 273 segment drops recorded in the case of the "Full RTS/CTS" method and only 146 DATA and ACK bearing MAC frame collisions - many of the drops would be explained in terms of RTS/CTS exchange failure.

In summary, this section has demonstrated that in the special case of the 5-node string topology, the increased spatial contention due to the RTS/CTS exchange results in a goodput penalty for the TCP agent. For the first time, spatial contention has been quantified in this special case, in terms of frame types (RTS/CTS, TCP DATA and ACK frames). It has been shown that an increase in collisions during the RTS/CTS exchange leads to increased segment drops and a decrease in achieved goodput. This example has also shown for the first time that a more conservative approach in the generation of RTS/CTS segments, through the "Partial RTS/CTS" method, results in goodput improvement. The next section, examines the implications of employing such RTS/CTS methods in several string and mesh topology settings and in a variety of ACK-thinning techniques. The overall aim is to investigate if the noted performance improvement in this case will be applicable in a more generalised setting.

5.5   Performance evaluation

This section expands on the scope of the previous examination of a specific string topology. The intent of this enquiry is to determine the merits in terms of goodput of the "Partial" and "No RTS/CTS" technique with respect to two ACK-thinning techniques in MANETs, delayed ACKs and DAA (as outlined in Sections 5.2.1 and 5.2.2 respectively). The dimension in focus for this performance analysis is, thus, the effects of each RTS/CTS strategy *per ACK-thinning method.*

5.5.1   Simulation setup

As with the analysis of the previous section, the same metrics of spatial contention are used in the following performance evaluation, namely the total number of frame collisions (COL_{TOTAL}) and the total number of frame drops due to repeated failed MAC layer transmissions ($FAIL_{TOTAL}$). Further the achieved goodput of TCP is recorded.

The topology setup of the simulation is outlined separately in each of the sections below. The simulation parameters, common to the performance evaluation conducted are outlined in Table 5.5 and are in agreement with previous ACK-thinning research work [6, 31]. Considerations on the ACK timer granularity are taken into account in this study in the manner outlined in Section 5.3.1.

5.5.2   Evaluation on string topologies

String topologies as commonly used to evaluate TCP performance in the presence of spatial contention [6, 26, 31, 43, 69, 105], are the focal point of interest in this section. Similarly to previous sections in this chapter, an FTP connection is set up between the endpoints of the string topology and runs throughout the simulation time before results are collected. The string topologies considered are of size n, where $4 \leq n \leq 12$; these are illustrated shown in a general form in Figure 5.2.

For this enquiry three ACK-handling paradigms have been evaluated with respect to achieved goodput and noted collisions and drops. These paradigms are a plain TCP

Table 5.5. Common Simulation Parameters

Parameter	Value
Channel Bandwidth	2Mbps
Signal Propagation	Two-Ray Ground
Packet Size	1460 bytes
TCP Agent	NewReno
ACK granularity response	10ms (Linux kernel 2.4)
Routing Protocol	AODV
Simulation Time	300 secs

receiver (plain ACKs), as the base case; delayed ACKs, as a popular optimisation of TCP [5], and the Dynamic Adaptive ACK strategy [31].

The goodput results for a single connection may be considered, with respect to the effects of the RTS/CTS strategy used on each ACK strategy. Figure 5.8 shows the goodput results of a plain TCP receiver on topologies of increasing hop-count for each of the RTS/CTS strategies examined here. The achieved goodput regardless of the ACK strategy used decreases as the hop-count (number of nodes in the string topology) increases, hence altering the ACK strategy does not alter that TCP behavioural characteristic as noted in the case of the "Full RTS/CTS" mechanism in previous research work [26, 43, 107, 108]. However, notably, goodput increases both when the "Partial RTS/CTS" and "No RTS/CTS" techniques are in use. In the case of the "Partial RTS/CTS" mechanism the increase is in the range of 6-9% for plain ACKs (Figure 5.7(a)), 3-6% for delayed ACKs (Figure 5.7(b)) and approx. 3% for Dynamic Adaptive ACKs (Figure 5.7(c)). The improvement in the case of disabling the RTS/CTS exchange is high and ranges within 17-23% for plain ACKs, 15-22% for delayed ACKs and 9-19% for Dynamic Adaptive ACKs. As ACK-thinning techniques are employed, the effectiveness of both the "Partial" and "No RTS/CTS" strategies diminishes, most notably for the "Partial" RTS/CTS technique. This is because ACK-thinning alleviates some of the spatial contention and there is less scope for improvement by this MAC layer modification.

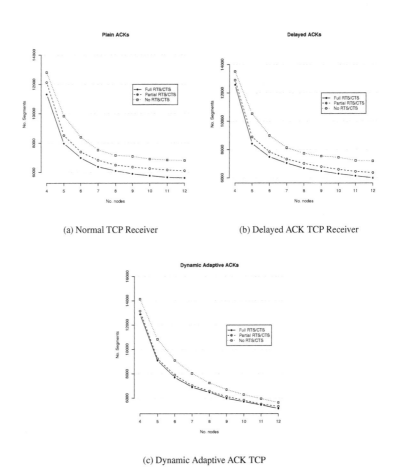

(a) Normal TCP Receiver (b) Delayed ACK TCP Receiver

(c) Dynamic Adaptive ACK TCP

Figure 5.8. Goodput against number of nodes in string topologies for a single TCP connection

Figure 5.9 presents the number of total collisions recorded for each strategy in the same scenarios. In all the RTS/CTS techniques as the number of nodes in the string increases so does spatial contention (indicated by the increasing number of frame collisions). This fact is reflected on the declining goodput as the string topology length increases (Figure 5.8). Especially in the case of the Dynamic Adaptive ACKs, the level of spatial contention is substantially less than the other two strategies, as there are no noted RTS/CTS exchanges and the TCP agent maintains a low congestion window (up to 4 segments), i.e. maintains few segments in flight. This results in few collisions as segments are few and tend to "spread" along the string.

We have also noted the results of the above scenario in the case of multiple TCP connections among the end-points. In this case, the *aggregate* goodput is considered. The results verify observations made above and the relevant goodput graphs may be found in Figure 5.10 and Figure 5.11 for 2 and 3 TCP connections respectively. It is worthy of mention that the goodput advantage of both the "Partial" and "No RTS/CTS" methods against the "Full RTS/CTS" exchange strategy remains consistent as more connections are employed on string topologies of the length used here. In particular, Table 5.6 presents the goodput performance improvement noted by employing the two RTS/CTS strategies against the full RTS/CTS exchange for the three types of ACK response methods (plain, delayed and Dynamic Adaptive ACKs).

For a single connection, generally, disabling RTS/CTS decreases the number of collisions compared to the other two strategies, but as the hop count increases, this trend does not hold consistently across methods, notably for the delayed ACKs technique for 9, 10 and 11 hops (Figure 5.9(b)) and the plain ACKs for 11 hops (Figure 5.9(a)). The overall results are presented in Figure 5.9. For more than one connection, disabling RTS/CTS consistently reduces the number of collisions throughout different hop counts. The relevant graphs depicting this for 2 and 3 TCP connections are included in Figure 5.10 and Figure 5.11 respectively. It can therefore be deduced that in these cases disabling the exchange leads to reduction in *spatial contention*, which is in turn reflected in the goodput results.

The last metric appraised for this section is the number of drops registered due to *repeated failed transmissions*. Figure 5.12 depicts the recorded number of total drops (including both TCP DATA and ACK segments) for a single TCP connection for the three ACK strategies. Both disabling RTS/CTS for ACK segments ("Partial RTS/CTS"

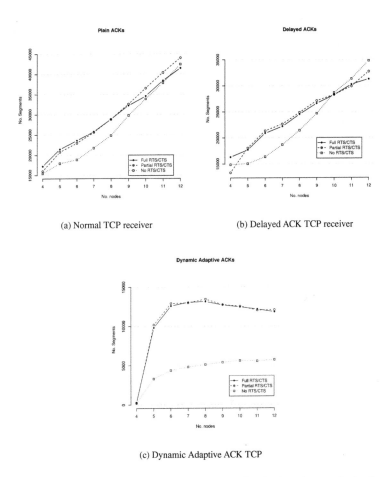

(a) Normal TCP receiver   (b) Delayed ACK TCP receiver

(c) Dynamic Adaptive ACK TCP

Figure 5.9. Number of collisions against number of nodes in string topologies for a single TCP connection

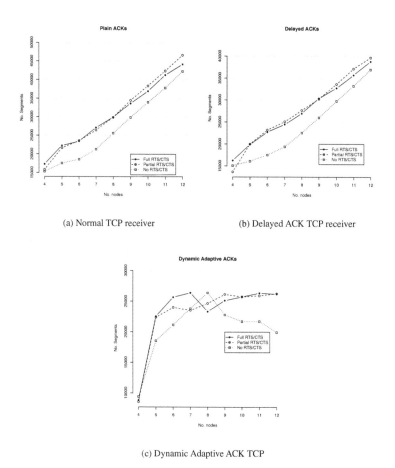

(a) Normal TCP receiver

(b) Delayed ACK TCP receiver

(c) Dynamic Adaptive ACK TCP

Figure 5.10. Number of collisions against number of nodes in string topologies for two TCP connections

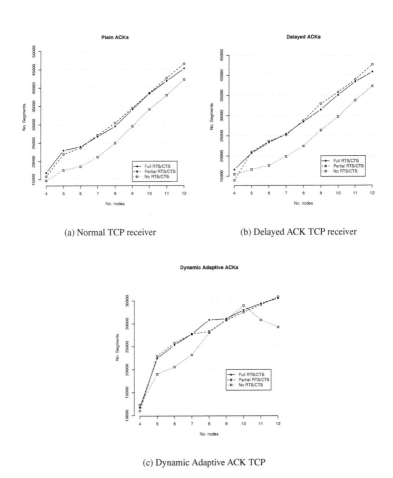

(a) Normal TCP receiver   (b) Delayed ACK TCP receiver

(c) Dynamic Adaptive ACK TCP

Figure 5.11. Number of collisions against number of nodes in string topologies for three TCP connections

Table 5.6. Range of goodput difference for each RTS/CTS strategy vs the "Full RTS/CTS" exchange for increasing number of TCP connections

ACK strategy	Range of goodput difference against "Full RTS/CTS"					
	2 TCP con.		3 TCP con.		4 TCP con.	
	PARTIAL	NO	PARTIAL	NO	PARTIAL	NO
Plain ACKs	7-8%	19-23%	6-7%	20-27%	5-8%	20-27%
Delayed ACKs	4-6%	19-22%	4-6%	19-24%	4-7%	20-28%
Dyn. Ad. ACKs	3-9%	8-19%	4-10%	8-19%	4-7%	8-19%
	5 TCP connections			6 TCP connections		
	PARTIAL		NO	PARTIAL		NO
Plain ACKs	6-8%		20-28%	6-8%		21-28%
Delayed ACKs	4-6%		21-27%	4-6%		22-28%
Dyn. Ad. ACKs	5-7%		8-21%	4-7%		10-23%

method) and disabling the exchange completely ("No RTS/CTS") result in fewer drops than the "Full RTS/CTS" method by a margin of 16-55% for the former and 19-50% for the latter when no ACK optimisations are employed (Figure 5.12(a)). Notably, as ACK optimisation methods are utilised, the number of recorded drops decreases as the ability of the MAC layer to cope with spatial contention improves. When delayed ACKs are used, as shown in Figure 5.12(b) there are notably less drops in the case of employing "Partial RTS/CTS" compared to the "Full" method, but disabling RTS/CTS altogether leads to higher segment drops in many instances (string topologies of 5,6,9 and 12 nodes in Figure 5.12(b)). This discrepancy reveals that not all "final" drops have an equal impact on goodput, i.e. certain segment drops are more damaging to goodput than others (note that in all cases, the goodput record for the "Full RTS/CTS" technique is less than the one recorded for the other methods).

The above observation may be explained once the *nature* of TCP segment loss is examined. In the case of a 5-node string topology in Figure 5.12(b) a total of 643 segment losses are noted for the "No RTS/CTS" strategy and 550 for "Full RTS/CTS" method. It would therefore seem that the latter handles spatial contention better than

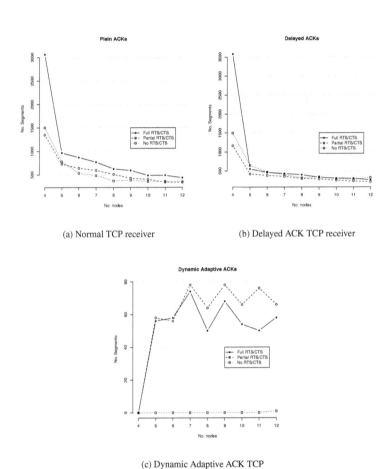

(a) Normal TCP receiver   (b) Delayed ACK TCP receiver

(c) Dynamic Adaptive ACK TCP

Figure 5.12. Number of drops due to consecutive failed transmissions against number of nodes in string topologies for a single TCP connection

Table 5.7. Detailed breakdown of segment loss due to repeated failed transmissions in a 4-hop string topology using a single TCP connection and delayed ACKs

RTS strategy	$FAIL_{RTS/CTS}$	$FAIL_{DATA}$	$FAIL_{ACK}$	$FAIL_{TOTAL}$
No RTS/CTS	0	35	608	643
Full RTS/CTS	275	201	74	550

Table 5.8. Detailed breakdown of segment loss due to repeated failed transmissions in a 5-hop string topology using one TCP connection

ACK strategy	$FAIL_{RTS/CTS}$	$FAIL_{DATA}$	$FAIL_{ACK}$	$textFAIL_{TOTAL}$
No RTS/CTS	0	37	427	464
Full RTS/CTS	228	168	60	456

the former; an observation not reflected in goodput as the "Full RTS/CTS" technique transfers, in total, 25% less segments. A breakdown of these losses reveals that most losses in the case of "No RTS/CTS" are ACK segments (608 out of 643) whilst the "Full RTS/CTS" records only 74 such losses. Further, the "No RTS/CTS" method experiences 35 TCP DATA drops as opposed to 201 for "Full RTS/CTS". Table 5.7 contains the complete data on the types of loss.

Intuitively, TCP DATA losses have a greater impact on goodput than ACK losses. Due to their cumulative nature, an ACK loss may be inconsequential if a subsequent ACK is received in time, i.e. before an RTO timeout is registered. For such an effect to occur, the average congestion window (cwnd) has to be sufficiently large so that several segments in the pipe would trigger ACK responses, some of which might be lost, but some of which would be received *in time so as not to trigger an RTO*. In the case of "No RTS/CTS" such a condition exists as the average value of cwnd is noted at 6 segments. Hence, it is the *nature* of segment loss which affects goodput in this case in tandem with the *amount* of segment loss. This statement holds true in all the other cases where the discrepancy occurs (6, 9, 12 node-string topologies). The corresponding losses breakdown in those cases are included in Tables 5.8, 5.9 and 5.10.

Table 5.9. Detailed breakdown of segment loss due to repeated failed transmissions in a 8-hop string topology using one TCP connection

ACK strategy	$\mathbf{FAIL}_{RTS/CTS}$	$\mathbf{FAIL}_{DATA}$	$\mathbf{FAIL}_{ACK}$	$textFAIL_{TOTAL}$
No RTS/CTS	0	17	322	339
Full RTS/CTS	164	113	51	328

Table 5.10. Detailed breakdown of segment loss due to repeated failed transmissions in a 11-hop string topology using one TCP connection

ACK strategy	$\mathbf{FAIL}_{RTS/CTS}$	$\mathbf{FAIL}_{DATA}$	$\mathbf{FAIL}_{ACK}$	$textFAIL_{TOTAL}$
No RTS/CTS	0	9	311	320
Full RTS/CTS	127	91	36	254

Two more points are worthy of note. Firstly that the Dynamic Adaptive ACK method significantly reduces the number of segment losses in all cases, as noted in previous research work [31], and especially in the case of a single TCP connection, eliminates them altogether when using the "NO RTS/CTS method", as shown in Figure 5.12. Secondly, it should be noted that the relevant segment drop results for 2 and 3 TCP connections follow the same trend as those for a single connection, with the added note that the inverse relationship between total number of drops and total achieved goodput holds in every case. The relevant segment drop results for 2 and 3 TCP connections are shown in Figures 5.13 and 5.14 whilst goodput is shown in Figures 5.15 and 5.15.

In conclusion, this section has examined the effect of the three RTS/CTS strategies on two popular ACK-thinning optimisations, namely delayed Acknowledgements and the Dynamic Adaptive Acknowledgements method on string topologies. The results indicate a substantial improvement in goodput in using either technique in all cases, with disabling the RTS/CTS exchange having the greatest impact in every case. The results have been explained in the context of frame losses (RTS/CTS or ACK/DATA TCP payloads) and using the two metrics introduced in the previous section. The next

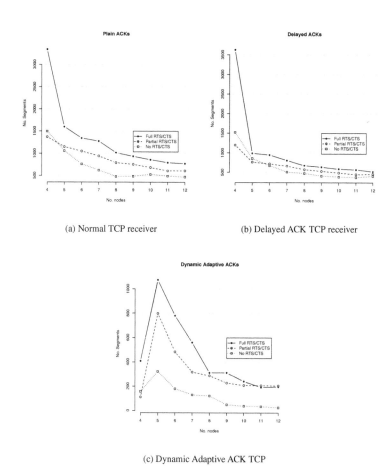

(a) Normal TCP receiver   (b) Delayed ACK TCP receiver

(c) Dynamic Adaptive ACK TCP

Figure 5.13. Number of drops due to consecutive failed transmissions against number of nodes in string topologies for two TCP connections

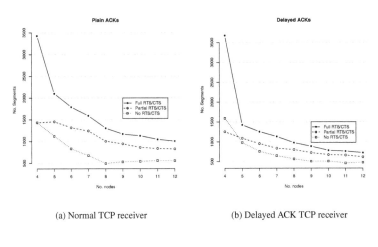

(a) Normal TCP receiver   (b) Delayed ACK TCP receiver

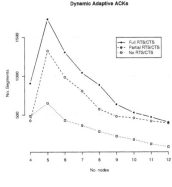

(c) Dynamic Adaptive ACK TCP

Figure 5.14. Number of drops due to consecutive failed transmissions against number of nodes in string topologies for three TCP connections

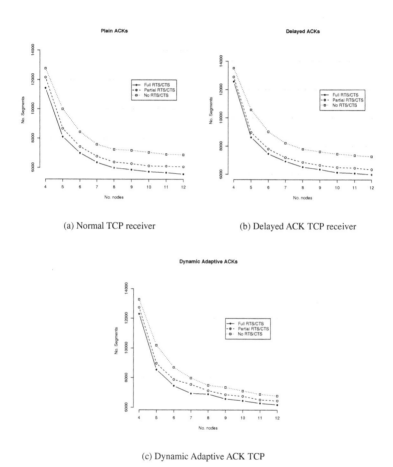

(a) Normal TCP receiver   (b) Delayed ACK TCP receiver

(c) Dynamic Adaptive ACK TCP

Figure 5.15. Goodput against number of nodes in string topologies for two TCP connections

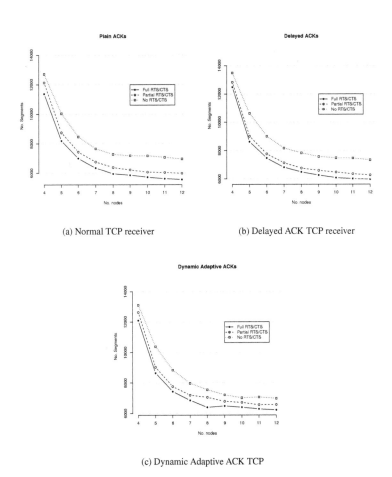

(a) Normal TCP receiver　　　　　(b) Delayed ACK TCP receiver

(c) Dynamic Adaptive ACK TCP

Figure 5.16. Goodput against number of nodes in string topologies for three TCP connections

Table 5.11. Goodput results for mesh topology with 3 TCP flows

ACK Strategy	Goodput Achieved		
	Full RTS/CTS	**Partial RTS/CTS**	**No RTS/CTS**
Plain ACKs	15953	16683(5%)	18798(18%)
Delayed ACKs	16115	1676(4%)	20968(30%)
Dyn. Ad. ACKs	16456	17260(5%)	19416(18%)

section examines the RTS/CTS handling methods with respect to intra-flow interference in a mesh topology as done previously in the literature [31, 43].

5.5.3   Evaluation on a mesh topology

The mesh topology as used in this section has been commonly used in literature to examine spatial contention and its effect on TCP when multiple interfering flows are present [31, 43, 104]. The focus of the simulation experiments is to identify whether the RTS/CTS exchange strategies affect throughput with respect to the ACK-thinning methods employed. As such, the scope of the investigation is similar to the previous section. However, the different topology setting offers insight into the interaction of TCP with the MAC layer mechanisms in the case of moderate *inter-flow* interference.

For the purposes of the following evaluation and discussion, the topology setup is set as shown in Figure 5.17. The setup involves a $5x5$ (25-node) mesh topology, where the horizontal and vertical distance of successive nodes is set to 200m. This setup mirrors that of previous work [31]. Two separate simulation scenarios are considered. First, three horizontal flows are active, as denoted by the solid arrows in Figure 5.17. This configuration offers *inter-flow spatial interference* alone; the flows do not share any common path but still interfere with each other due to the discrepancy between the interference and transmission ranges of their transceivers as described in Section 1.1.1. The second simulation scenario involves three additional vertical flows as depicted by the dashed arrows in Figure 5.17. This setup allows for both *buffer space* and *spatial* sharing between the flows; each flow shares its source and destination with another as shown in Figure 5.17.

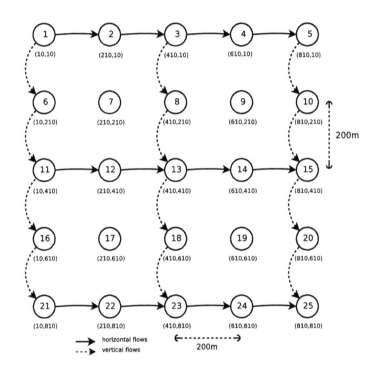

Figure 5.17. 25-node mesh topology

Table 5.12. Goodput results for mesh topology with 6 TCP flows

ACK Strategy	Goodput Achieved		
	Full RTS/CTS	**Partial RTS/CTS**	**No RTS/CTS**
Plain ACKs	16884	17879(6%)	23185(37%)
Delayed ACKs	16676	17171(3%)	21822(30%)
Dyn. Ad. ACKs	15848	19238(21%)	20601(30%)

Table 5.11 contains the goodput results for the plain, delayed and Dynamic Adaptive TCP ACK strategies for the different RTS/CTS methods, when 3 TCP flows are used. The values in parenthesis next to the numerical values for the "Partial" and "No RTS/CTS" methods indicate the performance improvement compared to the "Full RTS/CTS" method. For all three ACK strategies using an alternate strategy to the "Full RTS/CTS" exchange results in substantial improvement in goodput. As in the case of the string topologies, the "No RTS/CTS" method yields greater goodput improvement (18-30%) compared to the "Partial RTS/CTS" method (4-5%). In the case of 6 TCP flows, as shown in Table 5.12, the same observation holds true. It can be deduced that in both cases, i.e. whether spatial or buffer contention at the forwarding nodes is evident or otherwise, the alternative RTS/CTS strategies are beneficial goodput-wise compared to the "Full RTS/CTS" paradigm.

Table 5.13 presents the number of frame collisions noted for the mesh topology scenario using three TCP flows. It is noteworthy that the number of total collisions decreases for both the "Partial" and "No RTS/CTS" strategies, with the latter registering a more noteworthy decrease (12-35% as opposed to 1-19%). These improvements mirror the improvement noted in the case of multiple TCP connections in string topologies as discussed in the previous section. The equivalent results for the cross-traffic pattern of 6 TCP flows demonstrate show a similar trend and can be found in Table 5.14. Overall, the reduction in collisions indicate that spatial contention is reduced in this case, particularly when the RTS/CTS mechanism is disabled.

As indicated by the results in the string topology and the simulation trace examination in Section 5.4, the number of drops due to repeated failed transmissions is a useful indicator of the ability of the MAC mechanism to deal with spatial contention, i.e. to

Table 5.13. Collisions recorded for the mesh topology using 3 flows

ACK Strategy	Collisions Recorded		
	Full RTS/CTS	**Partial RTS/CTS**	**No RTS/CTS**
Plain ACKs	44155	43391(2%)	38905(12%)
Delayed ACKs	38593	38383(1%)	30312(21%)
Dyn. Ad. ACKs	36889	29564(19%)	23990(35%)

Table 5.14. Collisions recorded for the mesh topology using 6 flows

ACK Strategy	Collisions Recorded		
	Full RTS/CTS	**Partial RTS/CTS**	**No RTS/CTS**
Plain ACKs	57080	55964(2%)	43629(23%)
Delayed ACKs	50492	49290(3%)	41806(17%)
Dyn. Ad. ACKs	57080	55964(2%)	43629(24%)

effectively coordinate transmissions. In the case of 3 TCP flows the number of such recorded drops is substantially decreased for both the RTS/CTS strategies. The results may be found in Table 5.15. The percentage of reduction in drops, is consistently high regardless of the ACK strategy used, with an exception in the case of delayed ACKs. In that case, the "No RTS/CTS" registers only a 6% decrease. However, it should be noted that in that case the vast majority of drops (1251 out of 1325) are frames bearing an ACK payload, which has a smaller impact in goodput than the loss of TCP DATA segments. This phenomenon has been accounted for in the case of the string topologies in the previous section. Note that for the other two RTS/CTS techniques, that is the "Full" and "Partial RTS/CTS" methods, in the case of the delayed ACKs strategy, DATA-bearing frame losses severely dominate ACK-bearing ones (433 vs 217 in the case of "Partial RTS/CTS" and 429 vs 275 in the case of "Full RTS/CTS"). This fact is reflected in the recorded goodput for each method. When 6 TCP flows are employed the reduction in drops noted when using the RTS/CTS handshake alternatives as opposed to "Full RTS/CTS" is consistent across ACK strategies as can be seen in Table 5.16.

Table 5.15. Drops recorded for the mesh topology using 3 flows

ACK Strategy	Drops Recorded		
	Full RTS/CTS	Partial RTS/CTS	No RTS/CTS
Plain ACKs	2237	1596(28%)	1772(20%)
Delayed ACKs	1413	1083(23%)	1325(6%)
Dyn. Ad. ACKs	711	534(24%)	224(68%)

Table 5.16. Drops recorded for the mesh topology using 6 flows

ACK Strategy	Drops Recorded		
	Full RTS/CTS	Partial RTS/CTS	No RTS/CTS
Plain ACKs	2762	2363(14%)	1122(59%)
Delayed ACKs	2122	1846(13%)	920(56%)
Dyn. Ad. ACKs	2186	1473(32%)	424(80%)

In summary, in the case of the mesh topology examined employing either the "Partial" or "No RTS/CTS" strategies has a positive impact in goodput as either method helps *alleviate* and *handle* spatial contention, compared to the "Full RTS/CTS" handshake. It has been shown that this observation remains valid regardless of the ACK strategy employed.

5.6   Conclusions

This chapter has investigated the impact of introducing 802.11-compliant MAC layer optimisations on a TCP agent making use of ACK-thinning strategies in MANETs. The examination has considered both a traditional ACK-thinning paradigm used widely in wired networks, namely delayed ACKs and the recently introduced, MANET-specific Dynamic Adaptive ACKs method.

As a preamble to the performance evaluation, issues with respect to existing evaluation of ACK-thinning techniques were inspected and for the first time, otherwise implicit or hidden simulation parameters were made explicit. In particular, the feasible granularity of ACK-responses has been identified so as to ensure that subsequent simulation analysis corresponds to an implementable system. Further, the average path length produced by a popular mobility model in the literature (and throughout this study) has been examined. This analysis has identified the hop-count range of interest for the string topology simulation as utilised in the subsequent performance analysis. Finally, a MAC layer mechanism as employed in previous research studies has been explicitly examined with respect to other modes of operation, previously ignored in the literature. Specifically, an optimisation in the RTS/CTS exchange function of the 802.11 specification has been determined to be usable in tandem with plain and ACK-thinning enabled TCP agents.

Having identified different modes of the 802.11 MAC operation previously unexamined in the literature, this study has examined their effect on ACK responses in a special case of the string topology. First, two new metrics were introduced, measuring both the amount of spatial contention on TCP data exchanges and the ability of the MAC mechanism to effectively coordinate transmissions (maximise spatial reuse). Using these metrics and with the aid of simulation trace analysis, the alternative RTS/CTS functions have been shown to improve on spatial reuse.

This study has furthermore employed the new metrics and MAC layer optimisations to identify the latter's effect on TCP goodput in the general case of the string topology. The RTS/CTS optimisations were employed in tandem with ACK-thinning strategies to appraise the net throughput gain. The evaluation results verified that both spatial contention and the ability of the MAC mechanism to deal with it can improve with the deployment of the RTS/CTS alternative behaviour. Further, TCP goodput has been shown to improve substantially in the case of a plain TCP agent (up to 28%) as well as when delayed or Dynamic Adaptive ACKs are deployed (up to 28% and 23% respectively). The throughput gain has been shown to be consistent for single and multiple TCP connections.

In order to examine the effect of the RTS/CTS optimisations in the presence of inter-flow interference, simulations on a mesh topology have also been conducted. The simulation results have shown that the RTS/CTS optimisations at the MAC layer have a significant positive effect on TCP, resulting in improved goodput (up to 37%) for a plain TCP receiver. The improvement has been shown to exist for both delayed or Dynamic Adaptive enabled ACK receivers (up to 30% in both cases). The flow patterns under investigation have included scenarios of both isolated and path-sharing flows.

This chapter has been complementary to the previous one as it has examined mechanisms to improve TCP performance from the TCP receiver's perspective. As such, the two chapters may be viewed in conjunction with the prospect of combining their orthogonal suggestions to note cumulative goodput gains.

Chapter 6

Conclusions and future directions

6.1 Introduction

Mobile *Ad hoc* Networks (MANETs) have enjoyed significant research attention in the last few years as the increased popularity of wireless devices has brought the promise of ubiquitous connectivity closer to fruition. MANETs are in many ways ideally suited to facilitate such all-encompassing communications by acting as standalone, spontaneous networks or as impromptu gateways offering access to the Internet via collaborating access points. The layered approach principle in networking implies that development in MANETs can leverage on the experience and solutions developed for wired or wireless infrastructure networks in order to achieve rapid and reliable deployment.

However, existing protocols and mechanisms devised for wired networks have been based on assumptions challenged in a MANET setting, mainly due to the wireless nature of the shared medium and the requirement for mobility. Such discrepancies may lead to unpredictable behaviour and even a performance penalty in MANETs compared to their wired counterparts. The Transmission Control Protocol (TCP) is one such widely used mechanism, suited for reliable, end-to-end communications, which exhibits sub-optimal performance when used in a MANET. TCP has received early attention in the literature with respect to MANETs [13, 19, 24, 34, 35, 50], and its problems with regard to throughput in such an environment have been well documented [26, 40, 43, 105, 108].

Early work has been reported in the literature with respect to TCP performance in MANETs [19, 34]. However, as TCP has been shown to be greatly influenced by its interaction with the routing protocol [43, 107], such early studies did not account for performance issues with recent routing proposals. This dissertation has extended performance evaluation in the area by considering the goodput performance of popular TCP variants in concert with recently proposed routing protocols following the two trends in MANET routing, namely both proactive and reactive protocols.

Building on the results of the TCP performance analysis, we have investigated altering the congestion avoidance mechanism of TCP so as to improve goodput by making better use of the capacity of the wireless medium. The main consideration in the design of the mechanism has been ease of deployment, achievable by maintaining the end-to-end semantics of TCP and avoiding cross-layer dependencies. This work has shown that it is possible to introduce such changes and affect goodput positively without introducing overly intrusive changes to the TCP stack or significantly increasing the protocol's complexity.

As TCP conversations exhibit a sender/receiver dynamic with both ends being required to transmit through the wireless medium, our work has discussed and evaluated TCP optimisations for either perspective. Such considerations have led to the use and appraisal of an 802.11 optimisation at the MAC level which leads to better use of the wireless channel and an overall increase in TCP goodput. Throughout this work, special attention has been given to the explicit statement of assumptions made during simulation and, where appropriate, criticism has been offered to assumptions previously used in the literature.

6.2 Summary of contributions

This dissertation has focused on the examination of TCP behaviour in MANETs and has further introduced new mechanisms in the protocol stack aiming to enhance goodput in such environments. The major contributions made in this research work are summarised below.

- The first part of this dissertation has focused on examining the behaviour of TCP in a MANET setting. As a preliminary step to the actual TCP performance evaluation in dynamic topologies, simulation trace analysis is employed on a static topology over three popular routing protocols, namely *Ad hoc* On-Demand Distance Vector (AODV), Dynamic Source Routing (DSR) and Optimised Link State Routing (OLSR). The simulation trace analysis over AODV has revealed that buffering of segments at the routing layer can help avoid consecutive re-transmission timeouts (RTOs) and thus better utilise the wireless medium. Further, although previous research has identified the inability of the 802.11 protocol to coordinate transmissions in multihop networks (such as MANETs), this work has also confirmed this issue to exist in an ideal signal propagation setting where interference is not evident. Trace analysis of simulation over DSR has noted and demonstrated the positive effect on TCP goodput of the route caching and eaves-dropping functionality of the protocol. Finally, the analysis of TCP behaviour over OLSR has exposed its default routing parameters, as suggested in the RFC [28], to be sub-optimal with respect to TCP goodput for a small number of connections. We have offered an optimisation in the setting of the routing update interval which balances the trade-off between overhead and improved goodput performance when few TCP flows are present.

- TCP goodput has been examined in the context of dynamic topologies by taking into account limitations of the mobility model which have been resolved by recent research but not considered by previous performance evaluation studies. Four popular TCP variants, namely TCP Reno, NewReno, SACK and Vegas, have been considered in topologies depicting low, moderate and high mobility conditions. Overall, the results present a trend across routing protocols where TCP Vegas exhibits superior goodput over the reactive TCP variants especially under low mobility conditions. For instance, the difference in performance against Reno reaches 10% for AODV and OLSR and up to 12% for DSR. Through further careful tracing this performance advantage has been attributed to the less aggressive transmission policy of Vegas which leads to fewer segments in transit at any one time and thus reduces *spatial contention*. As such Vegas experiences less consecutive RTOs than the other variants and its goodput is not as severely affected.

- In view of the last observation above, the second part of the dissertation has examined methods of adopting a Vegas-like conservative sending rate for Reno-based TCP variants in order to improve goodput by mitigating spatial contention without compromising their reactive nature to packet loss. After considering changes in both the slow start and congestion avoidance phases of TCP, this work has proposed a slowdown parameter for reactive TCP variants as part of a method called Slow Congestion Avoidance (SCA). The proposed optimisation has been deliberately confined to the sending side of the TCP communicating pair and is implementable through simple alterations to the transport layer in order to facilitate ease of deployment. The new method has been evaluated in dynamic topologies and has been contrasted to another method existing in the literature, referred to as the adaptive Congestion Window Limit (CWL) technique. Simulations have shown that SCA can improve goodput by up to 20% over both standard TCP and the adaptive CWL variant in various mobility conditions.

- Further, this work has examined an adaption of the SCA method to improve TCP goodput in the case of multiple flows. Through simulation analysis an effective parameter for SCA has been determined and has been shown to improve goodput by up to 12% over plain TCP in dynamic topologies. Also, as the effectiveness of the slowdown parameter in SCA depends on the path length, we have considered utilising feedback from the routing protocol to adjust it dynamically. To this end, we have implemented and evaluated an adaptive SCA mechanism, which, however, does not lead to noticeable improvement over the standard SCA method.

- The third part of the dissertation has examined TCP optimisations for MANETs, applicable at the TCP receiver, which control the flow of ACK segments along the wireless path and are generally termed as *ACK-thinning* techniques. Through simulation examples it is noted that previous work in the literature has not taken a whole system view on the simulation parameters and has made assumptions which may not be applicable in practice. In particular, previous work has ignored the granularity requirements of the ACK responses and has not justified the choice of a hop-count range as the focal target for improvements in subsequent evaluation. Furthermore, other 802.11 MAC layer modes of operation have been largely ignored, even though they are part of the original protocol specification. This dissertation has shed more light on the issues relevant to the performance evaluation of ACK-thinning techniques and has also presented two 802.11 compliant optimisations adoptable at the MAC layer, which can improve TCP goodput.

• The new MAC level optimisations have been evaluated along with ACK-thinning techniques in a specific string topology scenario by introducing new metrics which help explain and quantify, in great detail, the level and causes of the observed goodput improvement. Finally, a broader scope of evaluation has been adopted where the new MAC optimisations have been appraised in a variety of string and mesh topologies. Notably, all recorded improvement has been accounted for and justified with the aid of the newly introduced metrics. In string topologies the goodput improvement exhibited by ACK-thinning methods using the MAC optimisations reaches 28% compared to employing the default MAC mechanism; the corresponding improvement in goodput in mesh topologies is 37%.

6.3 Directions for future work

In the course of this research and on reflection of the presented results, several prospects for future work are evident and some issues may be subject for further study. These are summarised below.

• The performance evaluation in Chapter 3 has compared four TCP variants popular in literature and commonly encountered in wired networks, namely TCP Reno, NewReno, SACK and Vegas. However, in recent times, there have been several other proposals in wired networks, which significantly alter TCP's congestion avoidance mechanism such as TCP Westwood [22, 45]. It would be an interest prospect to examine the impact of non-congestion related losses on those variants and note if the performance penalty incurred (if any) is of the same order as in the variants considered in this work.

• The random waypoint model was been used extensively in this dissertation to simulate mobility and its performance effect on TCP. Although this particular mobility model has been widely used in the literature there are several other models which account for different possible motion patterns. Considering the potentially ubiquitous nature of MANETs, such models could capture the mobility aspects of future MANET deployments and hence be representative of reality in some cases. A viable research prospect would be to examine TCP behaviour with regard to such other mobility models.

- In this work, the reaction of the congestion-aware reliable transport protocol (TCP) has been examined with relation to its application in a MANET setting. Recent research work, such as the Datagram Congestion Control Protocol (DCCP), has focused on introducing congestion awareness mechanisms to unreliable transport protocols, which exhibit TCP-friendliness [67]. A possible research avenue in the future could be the examination of the behaviour of such protocols in MANETs as these (similarly to TCP) are influenced by non-congestion related losses.

- A wide variety of mechanisms to enhance TCP throughput have been proposed in MANETs [34, 43, 104]. However, much of the subsequent evaluation has occurred in a homogeneous context, i.e. where implementations are well functioning and in agreement with each other. In reality, proposed alterations are deployed gradually in a network and communicating clients are expected to function adequately in a mixed environment. A research prospect along this lines would involve examining existing solutions in such heterogeneous settings and declaring whether gradual adoption is a viable option.

- The majority of research efforts with respect to MANETs have used simulation as a tool of extrapolating conclusions for issues under consideration. As in other research endeavours, simulation cannot (due to time and complexity considerations) predict results and provide insight for all possible scenarios. As such, a natural extension to the research efforts described in this dissertation would be to develop analytical models that can capture the performance behaviour of MANETs.

- There has been little activity in the deployment and performance measurement of actual MANET systems. Provided sufficient resources were available to materialise an actual MANET configuration, it would be useful to conduct measurements and verify simulations results reported in the literature. Apart from instilling confidence in existing work, such a deployment might reveal issues ignored in the simplifying assumptions of simulation or otherwise not captured in present simulation tools.

Appendix A

Simulation parameters

A.1   Routing agent parameters

Table A.1. AODV - Complete list of simulation parameters

Parameter	Value	Parameter	Value	Parameter	Value
Unidirectional Hack	OFF	Gratuitous RREQ	OFF	Expanding Ring Search	ON
Local Repair	ON	Receive n HELLOs	OFF	Jitter HELLOs	OFF
Wait on Reboot	OFF	Optimized HELLOs	OFF	Rate Limit	ON
LL Feedback	ON	Active Route Timeout	3 secs	TTL Start	2

Table A.2. DSR - Complete list of simulation parameters

Parameter	Value	Parameter	Value	Parameter	Value
Flush Link Cache	ON	Promiscuous Listening	ON	Broadcast Jitter	20ms
Route Cache Timeout	300	Send Buffer Timeout	30	Send Buffer Size	100
Request Table Size	64	Request Table IDs	16	Maximum Request Retransmissions	16
Maximum Request Period	10	Request Period	500	Non Propagation Request Timeout	30
Retransmission Buffer Size	50	Maintenance Holdoff Time	250	Maximum Maintenance Retransmissions	2
Network Layer ACKs	OFF	Use Passive ACKs	ON	Passive ACK Timeout	100
Gratuitous Reply Holdoff	ON	Maximum Salvage Count	15		

Table A.3. OLSR - Complete list of simulation parameters

Parameter	Value	Parameter	Value	Parameter	Value
Hello interval	1 sec	Refresh Interval	2 secs	TC Interval	5 secs
MID Interval	5 secs	HNA Interval	5 secs	Neighbourhood Hold Time	6 secs
Topology Hold Time	15 secs	Duplicate Hold Time	30 secs	MID Hold Time	15 secs
HNA Hold Time	15 secs	Max. Jitter	250ms	Willingness(WILL_DEFAULT)	3
TC Redundancy	OFF	MPR Coverage	1	Hysteresis Monitoring	OFF
Singal Moniroting	OFF	Delay Generation	OFF	Fast Route Calculation	OFF
Free Space Splitting Proportion Limit	OFF	Global Splitting Proportion Limit	0.5	Immediate Message Transmission	OFF

A.2   TCP agent parameters

Table A.4. TCP - Complete list of simulation parameters

Parameter	Value	Parameter	Value	Parameter	Value
Num. dupACKs	3	ECN	OFF	Timer Granularity	10ms
Max. RTO	60 secs	Min. RTO	200ms	FRTO	OFF
Delayed ACKs	ON	Max. Burst	3 segments	Lim. transmit	OFF
Vegas α	1	Vegas β	3	Vegas γ	1
No. SACK blocks	3	DSack generation	OFF	Seg. size	1460 bytes

Appendix B

Results supplement

B.1 Topology characteristics

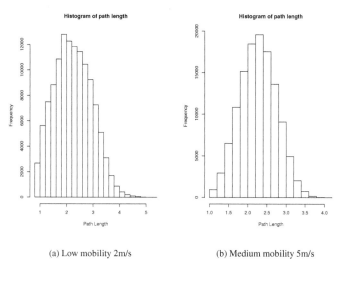

(a) Low mobility 2m/s   (b) Medium mobility 5m/s

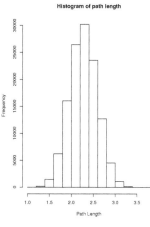

(c) High mobility 15m/s

Figure B.1. Histogram (Sturge's rule) of path lengths in a strip area (1500x300m)

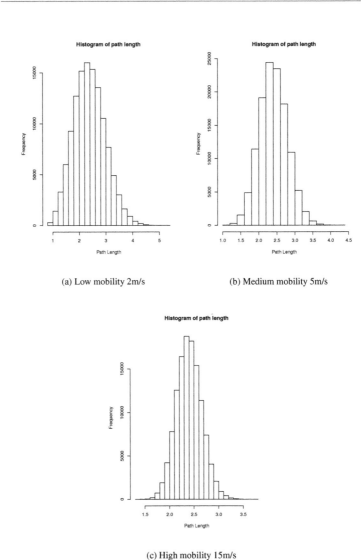

(a) Low mobility 2m/s   (b) Medium mobility 5m/s

(c) High mobility 15m/s

Figure B.2. Histogram (Sturge's rule) of path lengths in a strip area (1000x1000m)

B.2   SCA supplement

Table B.1. Default SCA Reno parameter on string topologies for AODV

Hop Count(h)	parameter	h	parameter
4	48	10	48
5	46	11	31
6	50	12	37
7	48	13	30
8	46	14	39
9	48	15	39

Appendix C

Publications during the course of this research

Book Chapters

- S. Papanastasiou, M. Ould-Khaoua, L. M. Mackenzie, *ACK-thinning techniques for TCP in MANETs*, under review for publication in a collection. Prof. A. Boukerche, Editor.

- S. Papanastasiou, M. Ould-Khaoua, L. M. MacKenzie, *A performance study of TCP in Mobile Ad Hoc Networks*. Performance Evaluation of Parallel, Distributed and Emergent Systems. (Volume 1 in Distributed, Cluster and Grid Computing), Nova Publishers, 2006.

- S. Papanastasiou, M. Ould-Khaoua, and L. M. MacKenzie. *TCP Developments in Mobile Ad hoc Networks*. Chapter 30, Handbook of Algorithms for Wireless and Mobile Networks and Computing. CRC Press, 2005.

Journal Papers

- S. Papanastasiou and M. Ould-Khaoua. *TCP Congestion window evolution and spatial reuse in MANETs*. Journal of Wireless Communications and Mobile Computing, 4(6):669-682, 2004.

- S. Papanastasiou and M. Ould-Khaoua. *Exploring the performance of TCP Vegas in Mobile Ad hoc Networks*. International Journal of Communication Systems, 17(2):163-177, 2004.

- M. Bani-Yassein, M. Ould-Khaoua, L. M. Mackenzie and S. Papanastasiou, *Performance Analysis of Adjusted Probabilistic Broadcasting in Mobile Ad Hoc Networks*, appearing in the International Journal of Wireless Information Networks, Pages 1-14, Springer Netherlands, Mar 2006.

Conference Papers

- S. Papanastasiou, V. Charissis, *Exploring the ad hoc network requirements of an automotive Head-Up Display interface.* In proceedings of Fifth International Symposium of Communication Systems, Networks and Digital Signal Processing, Patras, Greece, July 2006.

- S. Papanastasiou, L. Mackenzie, M. Ould-Khaoua, and V. Charissis. *On the interaction of TCP and Routing Protocols in MANETs.* In Proceedings of International Conference on Internet and Web Applications and Services/Advanced International Conference on Telecommunications (AICT-ICIW '06), pages 62-69, Guadeloupe, French Caribbean, February 2006. IEEE Computer Society Press.

- S. Papanastasiou, M. Ould-Khaoua, and L. M. Mackenzie. *On the evaluation of TCP in MANETs.* In Proceedings of International Workshop on Wireless Ad-hoc Networks (IWWAN 2005), London, United Kingdom, May 2005.

- S. H. A. Wahab, M. Ould-Khaoua, S. Papanastasiou, *Performance Analysis of the LWQ QoS Model in MANETs*, in proceedings of the 21st annual UK Performance Engineering Workshop (UKPEW '05), Newcastle, United Kingdom, July 14-15, 2005.

- S. Papanastasiou, M. Ould-Khaoua, and L. M. Mackenzie. *Exploring the effect of inter-flow interference on TCP performance in MANETs.* In Proceedings of Second International Working Conference of Performance Modelling and Evaluation of Heterogeneous Networks (HET-NETs 04), page P41. Bradford Print, 2004.

- S. Papanastasiou, L. M. MacKenzie, and M. Ould-Khaoua. *Reducing the degrading effect of hidden terminal interference in MANETs.* In Proceedings of the 7th ACM international symposium on Modeling, analysis and simulation of wireless and mobile systems (MSWiM), pages 311-314. ACM Press, 2004.

- S. Papanastasiou and M. Ould-Khaoua. *On the performance of TCP Vegas in Mobile Ad hoc Networks.* In Proceedings of the 2003 International Symposium on Performance Evaluation of Computer and Telecommunication Systems (SPECTS '03), pages 417-422, 2003.

Bibliography

[1] J. S. Ahn, P. B. Danzig, Z. Liu, and L. Yan. Evaluation of TCP Vegas: emulation and experiment. In *Proceedings of the conference on Applications, technologies, architectures, and protocols for computer communication*, pages 185–195. ACM Press, 1995.

[2] A. Ahuja, S. Agarwal, J. P. Singh, and R. Shorey. Performance of TCP over different routing protocols in mobile ad-hoc networks. In *Proceedings of IEEE Vehicular Technology Conference (VTC 2000)*, pages 2315–2319, May 2000.

[3] M. Allman. On the generation and use of TCP acknowledgments. *SIGCOMM Computer Communication Review*, 28(5):4–21, 1998.

[4] M. Allman, S. Floyd, and A. Medina. Measuring the evolution of transport protocols in the internet. http://www.icir.org/tbit/TCPevolution-Dec2004.pdf.

[5] M. Allman, V. Paxson, and W. Stevens. *TCP Congestion Control*. Internet Draft, http://www.ietf.org/rfc/rfc2581.txt, April 1999. Request For Comments.

[6] E. Altman and T. Jiménez. Novel Delayed ACK Techniques for Improving TCP Performance in Multihop Wireless Networks. In *Personal Wireless Communications*, volume 2775, pages 237–250. Springer-Verlag Heidelberg, September 2003.

[7] G. Anastasi, E. Borgia, M. Conti, and E. Grego. IEEE 802.11 Ad Hoc Networks: Performance Measurements. In *Proceedings of the 23rd International Conference on Distributed Computing Systems Workshops (ICDCSW'03)*, pages 758–763, May 2003.

[8] F. Anjum and L. Tassiulas. Comparative study of various TCP versions over a wireless link with correlated losses. *IEEE/ACM Transactions on Networking*, 11(3):370–383, 2003.

[9] I. Armuelles, H. Chaouchi, T. R. Valladares, I. Ganchev, M. O'Droma, and M. Siebert. On ad hoc networks in the 4g integration process. In *Proceedings of the third Annual Mediterranean Ad Hoc Networking Workshop (Med-Hoc 2004)*, June 2004.

[10] B. Bakshi, P. Krishna, N. Vaidya, and D. Pradhan. Improving performance of TCP over wireless networks. In *Proceedings of the 17th IEEE International Conference on Distributed Computer Systems (ICDSC)*, 1997.

[11] The FreeBSD Project. http://www.freebsd.org/.

[12] D. Bertsekas and R. Gallager. *Data Networks*, volume 1. Prentice Hall, 1987.

[13] S. Biaz and N. Vaidya. Discriminating congestion losses from wireless losses using interarrival times at the receiver. In *Proceedings of IEEE Symposium on Application-Specific Systems and Software Engineering and Technology (AS-SET'99)*, pages 10–17, March 1999.

[14] E. Blanton, M. Allman, K. Fall, and L. Wang. *A Conservative Selective Acknowledgment (SACK)-based Loss Recovery Algorithm for TCP*. Internet Draft, http://www.ietf.org/rfc/rfc3517.txt, April 2003. Experimental RFC.

[15] R. Braden. *Requirements for Internet Hosts - Communication Layers*. Internet Draft, http://www.ietf.org/rfc/rfc1122.txt, October 1989.

[16] L. Brakmo and L. Peterson. TCP Vegas: End to End Congestion Avoidance on a global Internet. *IEEE Journal on Selected Areas in Communications*, 13(8):1465–1480, October 1995.

[17] L. S. Brakmo, S. W. O'Malley, and L. L. Peterson. TCP Vegas: new techniques for congestion detection and avoidance. In *Proceedings of the conference on Communications architectures, protocols and applications*, pages 24–35. ACM Press, 1994.

[18] L. S. Brakmo and L. L. Peterson. Performance problems in BSD4.4 TCP. *SIG-COMM Computer Communications Review*, 25(5):69–86, 1995.

[19] J. Broch, D. A. Maltz, D. B. Johnson, Y.-C. Hu, and J. Jetcheva. A performance comparison of multi-hop wireless ad hoc network routing protocols. In *Proceedings of the fourth annual ACM/IEEE international conference on Mobile computing and networking*, pages 85–97. ACM Press, 1998.

[20] T. Camp, J. Boleng, and V. Davies. A Survey of Mobility Models for Ad Hoc Network Research. *Wireless Communication & Mobile Computing (WCMC)*, 2(5):483–502, 2002.

[21] J. Cartigny, D. Simplot, and I. Stojmenovic. Localized minimum-energy broadcasting in ad-hoc networks. In *Proceedings of Twenty-Second Annual Joint Conference of the IEEE Computer and Communications Societies (INFOCOM 2003)*, volume 3, March 2003.

[22] C. Casetti, M. Gerla, S. Mascolo, M. Y. Sanadidi, and R. Wang. TCP Westwood: end-to-end congestion control for wired/wireless networks. *Wireless Networks*, 8(5):467–479, 2002.

[23] R. Chandra, V. Bahl, and P. Bahl. Multinet: connecting to multiple ieee 802.11 networks using a single wireless card. In *Proceedings of Twenty-Third Annual Joint Conference of the IEEE Computer and Communications Societies (INFO-COM 2004)*, volume 2, pages 882 – 893, March 2004.

[24] K. Chandran, S. Raghunathan, S. Venkatesan, and R. Prakash. A feedback based scheme for improving TCP performance in ad-hoc wireless networks. In *Proceedings of the 18th annual International Conference on Distributed Computing Systems*, pages 472–479, May 1998.

[25] K. Chen, Y. Xue, and K. Nahrstedt. On Setting TCP's Congestion Window Limit in Mobile Ad Hoc Networks. In *Proceedings of the 38th annual IEEE International Conference on Communications ICC 2003*, pages 1080–1084, May 2003.

[26] K. Chen, Y. Xue, S. H. Shah, and K. Nahrstedt. Understanding bandwidth-delay product in mobile ad hoc networks. *Computer Communications*, 27(10):923–934, June 2004.

[27] D. D. Clark. *Window and acknowledgment strategy in TCP*. Internet Draft, http://www.ietf.org/rfc/rfc813.txt, July 1982.

[28] T. Clausen and P. Jacquet. *Optimized Link State Routing Protocol (OLSR)*. http://www.ietf.org/rfc/rfc3626.txt, October 2003. Experimental RFC.

[29] S. Corson and J. Macker. *Mobile Ad hoc Networking (MANET): Routing Protocol Performance Issues and Evaluation Considerations*. Internet Draft, http://www.ietf.org/rfc/rfc2501.txt, January 1999. Informational RFC.

[30] S. Corson and J. Macker. *Mobile Ad hoc Networking (MANET):Routing Protocol Performance Issues and Evaluation Considerations*. http://www.ietf.org/rfc/rfc2501.txt, January 1999. Informational RFC.

[31] R. de Oliveira and T. Braun. A Dynamic Adaptive Acknowledgment Strategy for TCP over Multihop Wireless Networks. In *Proceedings of Twenty-Fourth Annual Joint Conference of the IEEE Computer and Communications Societies (INFOCOM 2005)*, volume 3, March 2005.

[32] O. Dousse, F. Baccelli, and P. Thiran. Impact of interferences on connectivity in ad hoc networks. In *Twenty-Second Annual Joint Conference of the IEEE Computer and Communications Societies, INFOCOM 2003*, volume 3, pages 1724–1733. IEEE Computer Society Press, March 2003.

[33] R. Dube, C. Rais, K.-Y. Wang, and S. Tripathi. Signal stability-based adaptive routing (SSA) for ad hoc mobile networks. In *IEEE Personal Communications*, volume 4, pages 36–45, Feb 1997.

[34] T. D. Dyer and R. V. Boppana. A comparison of TCP performance over three routing protocols for mobile ad hoc networks. In *Proceedings of the 2001 ACM International Symposium on Mobile ad hoc networking & computing*, pages 56–66. ACM Press, 2001.

[35] H. Elaarag. Improving TCP performance over mobile networks. *ACM Computing Surveys (CSUR)*, 34(3):357–374, 2002.

[36] K. Fall and S. Floyd. Simulation-based comparisons of Tahoe, Reno and SACK TCP. *SIGCOMM Comput. Commun. Rev.*, 26(3):5–21, 1996.

[37] K. Fall and K. Varadhan. The ns manual - the VINT project. http://www.isi.edu/nsnam/ns/ns-documentation.html.

[38] S. Floyd, T. Henderson, and A. Gurtov. *The NewReno Modification to TCP's Fast Recovery Algorithm*. Internet Draft, http://www.ietf.org/rfc/rfc3782.txt, April 2004. Standards Track.

[39] S. Floyd and V. Jacobson. Traffic phase effects in packet-switched gateways. *SIGCOMM Computer Communications Review*, 21(2):26–42, 1991.

[40] S. Floyd and V. Jacobson. Random early detection gateways for congestion avoidance. *IEEE/ACM Transactions on Networking (TON)*, 1(4):397–413, 1993.

[41] S. Floyd, J. Mahdavi, M. Mathis, and M. Podolsky. *An Extension to the Selective Acknowledgement (SACK) Option for TCP*. Internet Draft, http://www.ietf.org/rfc/rfc2883.txt, July 2000. Experimental RFC.

[42] Z. Fu, X. Meng, and S. Lu. How Bad TCP Can Perform In Mobile Ad Hoc Networks. In *ISCC '02: Proceedings of the Seventh International Symposium on Computers and Communications (ISCC'02)*, page 298, Washington, DC, USA, 2002. IEEE Computer Society.

[43] Z. Fu, P. Zerfos, H. Luo, S. Lu, L. Zhang, and M. Gerla. The impact of multihop wireless channel on TCP throughput and loss. In *Proceedings of Twenty-Second Annual Joint Conference of the IEEE Computer and Communications Societies (INFOCOM 2003)*, volume 3, pages 1744–1753, March 2003.

[44] D. Goldsman. Simulation output analysis. In *WSC '92: Proceedings of the 24th conference on Winter simulation*, pages 97–103, New York, NY, USA, 1992. ACM Press.

[45] L. A. Grieco and S. Mascolo. Performance evaluation and comparison of west-wood+, new reno, and vegas tcp congestion control. *SIGCOMM Comput. Commun. Rev.*, 34(2):25–38, 2004.

[46] M. Gunes and D. Vlahovic. The performance of the TCP/RCWE enhancement for ad-hoc networks. In *Proceedings of the Seventh International Symposium on Computers and Communications (ISCC 2002)*, pages 43–48. IEEE Computer Society Press, July 2002.

[47] C. Hedrick. *Routing Information Protocol.* Internet Draft, http://www.ietf.org/rfc/rfc1058.txt, June 1988.

[48] High Performance Communications (HIPERCOM) Project. Object-Oriented Optimized Link State Routing Protocol (OOLSR). http://hipercom.inria.fr/OOLSR/.

[49] G. Holland and N. Vaidya. Analysis of TCP performance over mobile ad hoc networks. In *Proceedings of the fifth annual ACM/IEEE international conference on Mobile computing and networking*, pages 219–230. ACM Press, 1999.

[50] G. Holland and N. Vaidya. Analysis of TCP Performance over Mobile Ad Hoc Networks. Technical Report 99-004, Texas A&M University, February 1999.

[51] P.-H. Hsiao, H. T. Kung, and K.-S. Tan. Active Delay Control for TCP. In *Global Telecommunications Conference, 2001. GLOBECOM '01*, volume 3, pages 25–29. IEEE Computer Society Press, November 2001.

[52] IEEE Standards Association. IEEE P802.11, The Working Group for Wireless LANs. http://grouper.ieee.org/groups/802/11/index.html.

[53] IEEE Standards Association. IEEE P802.16, The Working Group for Broadband Wireless Access (BBWA). http://grouper.ieee.org/groups/802/16/index.html.

[54] U. o. S. C. Information Sciences Institute. *Transmission Control Protocol.* Internet Draft, http://www.ietf.org/rfc/rfc793.txt, September 1981. Request For Comments.

[55] A. J. Izenman. Recent developments in nonparametric density estimation. *Journal of the American Statistical Association*, 86(413):205–224, 1991.

[56] V. Jacobson. Congestion avoidance and control. *ACM SIGCOMM Computer Communication Review*, 18(4):314–329, 1988.

[57] S. Jin, L. Guo, I. Matta, and A. Bestavros. A spectrum of tcp-friendly window-based congestion control algorithms. *IEEE/ACM Trans. Netw.*, 11(3):341–355, 2003.

[58] P. Johansson, T. Larsson, N. Hedman, B. Mielczarek, and M. Degermark. Scenario-based performance analysis of routing protocols for mobile ad-hoc networks. In *Proceedings of the fifth annual ACM/IEEE international conference on Mobile computing and networking*, pages 195–206. ACM Press, 1999.

[59] D. B. Johnson and D. A. Maltz. *Dynamic Source Routing in Ad Hoc Wireless Network*, chapter 5, pages 153–181. Mobile Computing. Kluwer Academic Publishers, 1996.

[60] D. B. Johnson, D. A. Maltz, and Y.-C. Hu. *The Dynamic Source Routing Protocol for Mobile Ad Hoc Networks (DSR)*. Internet Draft, draft-ietf-manet-dsr-10.txt, July 2004. Work in progress.

[61] C. E. Jones, K. M. Sivalingam, P. Agrawal, and J. C. Chen. A Survey of Energy Efficient Network Protocols for Wireless Networks. *Wirel. Netw.*, 7(4):343–358, 2001.

[62] G. Judd and P. Steenkiste. Repeatable and realistic wireless experimentation through physical emulation. *SIGCOMM Comput. Commun. Rev.*, 34(1):63–68, 2004.

[63] E.-S. Jung and N. H. Vaidya. A power control MAC protocol for ad hoc networks. In *Proceedings of the eighth annual international conference on Mobile computing and networking*, pages 36–47. ACM Press, 2002.

[64] A. Kamerman and L. Monteban. WaveLAN II: A high-performance wireless LAN for unlicensed band. *Bell Labs Technical Journal*, 2(3):118–133, Summer 1997.

[65] T. K. Kanth, S. Ansari, and M. H. Mehkri. Performance enhancement of TCP on multihop ad hoc wireless networks. In *Proceedings of IEEE International Conference on Personal Wireless Communications*, pages 90–94, October 2002.

[66] D. Kim, C.-K. Toh, and Y. Choi. TCP-BuS: Improving TCP Performance in Wireless Ad Hoc Networks. *Journal of Communications and Networks*, 3(2):175–186, June 2001.

[67] E. Kohler, M. Handley, and S. Floyd. *Datagram Congestion Control Protocol*. Internet Draft, http://www.ietf.org/rfc/rfc4340.txt, March 2006. Standards Track.

[68] R. B. S. Konduru. An adaptive distance vector routing algorithm for mobile, ad hoc networks. In *Proceedings of Twentieth Annual Joint Conference of the IEEE Computer and Communications Societies (INFOCOM 2001)*, volume 3, pages 1753–1762, April 2001.

[69] S. Kopparty, S. Krishmniurthy, M. Faloutsos, and S. Tripathi. Split TCP for mobile ad hoc networks. In *Global Telecommunications Conference, 2002. GLOBECOM '02*, volume 1, pages 138–142. IEEE Computer Society Press, November 2002.

[70] H. Kung, K.-S. Tan, and P.-H. Hsiao. TCP with sender-based delay control. *Computer Communications*, 26(14):1614–1621, September 2003.

[71] S. Kurkowski, T. Camp, and M. Colagrosso. Manet simulation studies: the incredibles. *SIGMOBILE Mob. Comput. Commun. Rev.*, 9(4):50–61, 2005.

[72] Y.-C. Lai and C.-L. Yao. Performance comparison between TCP Reno and TCP Vegas. *Computer Communications*, 2002.

[73] Q. Li, J. Aslam, and D. Rus. Online power-aware routing in wireless Ad-hoc networks. In *Proceedings of the 7th annual international conference on Mobile computing and networking*, pages 97–107. ACM Press, 2001.

[74] H. Lim, K. Xu, and M. Gerla. TCP Performance over Multipath Routing in Mobile Ad Hoc Networks. In *Proceedings of the 38th annual IEEE International Conference on Communications ICC 2003*, pages 1064–1068, May 2003.

[75] G. Lin, G. Noubir, and R. Rajamaran. Mobility Models for Ad hoc Network Simulation. In *Proceedings of Twenty-Third Conference of the IEEE Communications Society (INFOCOM 2003)*, volume 1, pages 454–463, March 2004.

[76] The Linux Kernel Archives. http://www.kernel.org/.

[77] J. Liu and S. Singh. ATCP: TCP for Mobile Ad Hoc Networks. *IEEE Journal on Selected Areas in Communications*, 19(7):1300–1315, July 2001.

[78] R. Ludwig and K. Sklower. The Eifel retransmission timer. *SIGCOMM Computer Communications Review*, 30(3):17–27, 2000.

[79] H. Lundgren, E. Nordström, and C. Tschudin. Coping with communication gray zones in ieee 802.11b based ad hoc networks. In *WOWMOM '02: Proceedings of the 5th ACM international workshop on Wireless mobile multimedia*, pages 49–55. ACM Press, 2002.

[80] I. MANET. Mobile ad-hoc networks (manet) ietf working group. http://www.ietf.org/html.charters/manet-charter.html.

[81] M. Mathis, J. Mahdavi, S. Floyd, and A. Romanow. *TCP Selective Acknowledgment Options*. Internet Draft, http://www.ietf.org/rfc/rfc2018.txt, October 1996. Proposed Standard.

[82] T. W. Mehran Abolhasan and E. Dutkiewicz. A review of routing protocols for mobile ad hoc networks. *Ad Hoc Networks*, 2(1):1–22, January 2004.

[83] W. Navidi and T. Camp. Stationary distributions for the random waypoint mobility model. *IEEE Transactions on Mobile Computing*, 3(1):99–108, 2004.

[84] E. Nordström. DSR-UU: A Dynamic Source Routing protocol implementation. http://core.it.uu.se/AdHoc/DsrUUImpl.

[85] S. Papanastasiou, L. Mackenzie, M. Ould-Khaoua, and V. Charissis. On the interaction of TCP and Routing Protocols in MANETs. In *International Conference on Internet and Web Applications and Services/Advanced International Conference on Telecommunications (AICT-ICIW '06)*, pages 62–69, Guadeloupe, French Caribbean, February 2006. IEEE Computer Society Press.

[86] S. Papanastasiou and M. Ould-Khaoua. Exploring the performance of TCP Vegas in Mobile Ad hoc Networks. *International Journal of Communication Systems*, 17(2):163–177, 2004.

[87] S. Papanastasiou and M. Ould-Khaoua. TCP and Interference in Mobile Ad hoc Networks. Technical Report No. TR-2004-160, Department of Computing Science, University of Glasgow, February 2004.

[88] A. Patwardhan, J. Parker, A. Joshi, M. Iorga, and T. Karygiannis. Secure Routing and Intrusion Detection in Ad Hoc Networks. In *Proceedings of the 3rd IEEE International Conference on Pervasive Computing and Communications (PERCOM)*. IEEE, March 2005.

[89] V. Paxson and M. Allman. *Computing TCP's Retransmission Timer.* Internet Draft, http://www.ietf.org/rfc/rfc2988.txt, November 2000.

[90] V. Paxson, M. Allman, S. Dawson, W. Fenner, J. Griner, I. Heavens, K. Lahey, J. Semke, and B. Volz. *Known TCP Implementation Problems.* Internet Draft, http://www.ietf.org/rfc/rfc2525.txt, March 1999. Request For Comments.

[91] C. E. Perkins. *Ad Hoc Networking.* Addison Wesley Professional, 2001.

[92] C. E. Perkins, E. M. Belding-Royer, and S. R. Das. *Ad hoc On-Demand Distance Vector (AODV) Routing.* Request For Comments, http://www.ietf.org/rfc/rfc3561.txt, July 2003. Experimental RFC.

[93] C. E. Perkins and P. Bhagwat. Highly dynamic Destination-Sequenced Distance-Vector routing (DSDV) for mobile computers. In *Proceedings of the conference on Communications architectures, protocols and applications*, pages 234–244. ACM Press, 1994.

[94] T. Plesse, J. Lecomte, C. Adjih, M. Badel, and P. Jacquet. Olsr performance measurement in a military mobile ad-hoc network. In *Proceedings of the 24th International Conference on Distributed Computing Systems Workshops (ICDCSW'04)*, volume 6, pages 704–709. IEEE Computer Society, 2004.

[95] V. Ramarathinam and M. A. Labrador. Performance Analysis of TCP over Static Ad Hoc Wireless Networks. In *Proceedings of the 12th International Conference on Parallel and Distributed Computing Systems (PDCS-2000)*, pages 410–415. ACTA Press, 2000.

[96] R. Sollacher, M. Greiner, and I. Glauche. Impact of interference on the wireless ad-hoc networks capacity and topology. *Wireless Networks*, 12(1):53–61, 2006.

[97] W. Stevens. *TCP Slow Start, Congestion Avoidance, Fast Retransmit, and Fast Recovery Algorithms*. Internet Draft, http://www.ietf.org/rfc/rfc2001.txt, January 1997.

[98] W. R. Stevens. *TCP/IP Illustrated*, volume 1. Addison-Wesley, Reading, MA, 1994.

[99] M. Takai, J. Martin, and R. Bagrodia. Effects of wireless physical layer modeling in mobile ad hoc networks. In *MobiHoc '01: Proceedings of the 2nd ACM international symposium on Mobile ad hoc networking & computing*, pages 87–94, New York, NY, USA, 2001. ACM Press.

[100] Y.-C. Tseng, S.-Y. Ni, and E.-Y. Shih. Adaptive Approahes to Relieving Broadcast Storms in a Wireless Multihop Mobile Ad Hoc Network. In *ICDCS '01: Proceedings of the The 21st International Conference on Distributed Computing Systems*, pages 481–488, Washington, DC, USA, 2001. IEEE Computer Society.

[101] P. D. Welch. The statistical analysis of simulation results. In S. Lavenberg, editor, *The Computer Performance Modeling Handbook*, pages 268–328. Academic Press, 1983.

[102] H. Westman. *Reference Data for Radio Engineers*. Howard W. Sams Co., 6th edition edition, 1997.

[103] K. Xu, M. Gerla, and S. Bae. How effective is the IEEE 802.11 RTS/CTS handshake in ad hoc networks? In *Global Telecommunications Conference, 2002. GLOBECOM '02*, volume 1, pages 72–76, November 2002.

[104] K. Xu, M. Gerla, L. Qi, and Y. Shu. Enhancing TCP fairness in ad hoc wireless networks using neighborhood RED. In *Proceedings of the 9th annual international conference on Mobile computing and networking*, pages 16–28. ACM Press, 2003.

[105] S. Xu and T. Saadawi. Evaluation for TCP with delayed ACK option in wireless multi-hop networks. In *Proceedings of IEEE Vehicular Technology Conference (VTC 2001)*, volume 1, pages 267–271, 2001.

[106] S. Xu and T. Saadawi. Performance evaluation of TCP algorithms in multi-hop wireless packet networks. *Wireless Communications and Mobile Computing*, 2(1):85–100, March 2002.

[107] S. Xu and T. Saadawi. Revealing the problems with 802.11 medium access control protocol in multi-hop wireless ad hoc networks. *Computer Networks*, 38(4):531–548, March 2002.

[108] S. Xu, T. Saadawi, and M. Lee. Comparison of TCP Reno and Vegas in wireless mobile ad hoc networks. In *Proceedings of 25th Annual IEEE Conference on Local Computer Networks (LCN'00)*, pages 42–43, November 2000.

[109] E. L. Yan. Empirical Analyses of SACK TCP Reno and Modified TCP Vegas. http://citeseer.nj.nec.com/246505.html.

[110] J. Yoon, M. Liu, and B. Noble. Random waypoint considered harmful. In *Proceedings of Twenty-Second Annual Joint Conference of the IEEE Computer and Communications Societies (INFOCOM 2003)*, volume 2, pages 1312–1321, March 2003.

[111] X. Yu. Improving TCP performance over mobile ad hoc networks by exploiting cross-layer information awareness. In *MobiCom '04: Proceedings of the 10th annual international conference on Mobile computing and networking*, pages 231–244. ACM Press, 2004.

[112] T. Yuki, T. Yamamoto, M. Sugano, M. Murata, H. Miyahara, and T. Hatauchi. Performance Improvement of TCP over an Ad Hoc Network by Combining of Data and ACK Packets. In *The 5th Asia-Pacific Symposium on Information and*

Telecommunication Technologies (APSITT 2003), pages 339–344. IEEE Computer Society, November 2003.

www.ingramcontent.com/pod-product-compliance
Lightning Source LLC
LaVergne TN
LVHW022312060326
832902LV00020B/3418